Memoirs of the American Mathematical Society

Number 216

Alfred Magnus

Non-spherical principal series representations of a semisimple Lie group

Published by the

AMERICAN MATHEMATICAL SOCIETY

Providence, Rhode Island, USA

May 1979 · Volume 19 · Number 216 (first of 2 numbers)

Abstract

Non-spherical principal series representations of a real semisimple Lie group are studied. These are representations, induced by a one-dimensional representation of a minimal parabolic subgroup, which have a one-dimensional subspace left stable by a maximal compact subgroup of the original group G.

Necessary and sufficient conditions for such a representation to be irreducible, or to be cyclic, are found, in terms of parameters determined by certain rank one subgroups of G. A sufficient condition for such a representation to be unitary is found, and the condition is shown to be necessary in the rank one case.

The methods used in the proofs are in many cases similar to those of B. Kostant's "Irreducibility of Certain Series of Representations".

AMS(MOS) Classification (1970). 22E45

Library of Congress Cataloging in Publication Data

Magnus, Alfred, 1951-
Non-spherical principal series representations of a semisimple Lie group.

(Memoirs of the American Mathematical Society ; no. 216 ISSN 0065-9290)
Bibliography: p.
1. Lie groups. 2. Representations of groups. I. Title. II. Series: American Mathematical Society. Memoirs ; no. 216.
QA3.A56 no. 216 [QA387] 510'.8s [512'.55]
ISBN 0-8218-2216-0 79-10157

Table of Contents

Introduction

Certain non-spherical principal series representations of a semi-simple Lie group G are examined in this paper. Methods developed by Kostant in "Irreducibility of Certain Series of Representations", (4a.), henceforth abbreviated ICSR, for dealing with the spherical principal series are adapted to deal with another case. We find exactly which of these representations are infinitesimally irreducible, in terms of certain parameters depending on positive roots and the reductive subgroups of G corresponding to these roots, and find a sufficient condition for unitarity of the irreducible representations. This condition is shown to be necessary if the group has rank one.

We now give a short outline of the paper. Let G = KAN be the Iwasawa decomposition of a real semi-simple Lie group. Let τ be a character on K; thus, we assume K is not semisimple. The representations we consider are principal series representations which have a one-dimensional subspace on which K acts according to τ. By the Frobenius reciprocity theorem, one sees that these are representations induced from a character on MAN (M is the centralizer of A in K) whose restriction to M is $\tau|_M$, and whose restriction to A is an arbitrary character.

Now, G has Lie algebra $\underline{g} = \underline{k} + \underline{a} + \underline{n}$, and the principal series representation π^τ_λ, λ $\underline{a}'_C$, the complex dual of $\underline{a}$ (which indexes the characters on A) gives a representation on $U = U(\underline{g})$, the universal enveloping algebra of $\underline{g}$, on the space X^τ_λ of K-finite vectors. Most of the paper is devoted to investigating the action of U on X^τ_λ. Note that τ is a character on $\underline{k}$ by differentiation.

Let $\underline{k}^{\#} \subset U$ be the span of $\{X - \tau(X) \mid X \text{ in } \underline{k}\}$. We define a projection $u \to p^\tau_u$: $U \to U(\underline{a})$ by the direct sum decomposition $U = (\underline{k}^{\#}U + U\underline{n}) + U(\underline{a})$. This projection is used in ICSR for $\tau = 0$. Recalling the Cartan decomposition $\underline{g} = \underline{k} + \underline{p}$, and the symmetrization map β: $S(\underline{g}) \to U(\underline{g})$, a subspace $H^* \subset \beta(S(\underline{p}))$ is found in ICSR such that K acts on H^*, and $H^* = X^0_\lambda$ as a K-module.

Now, calling Γ the set of finite dimensional irreducible representations of K which occur in H^*, for each γ in Γ we define $U(\underline{a})$-valued matrices P^γ_τ and R^γ_τ, which are Kostant's P^γ and R^γ matrices if $\tau = 0$.

Essentially, the entries of P^γ_τ are of the form $p^\tau_{u_{ij}}$, where u_{ij} in H^*, and, considering $U(\underline{a})$ as the polynomial algebra on $\underline{a}'_C$, we essentially obtain, for λ in $\underline{a}'_C$, $R^\gamma_\tau(\lambda) = P^\gamma_\tau(\lambda)P^\gamma_\tau(-\lambda-2\rho)$. Here, ρ is one half the sum of the positive roots, and the principal series is indexed so that all the representations $\pi^\tau_{i\lambda-\rho}$ are unitary, for λ in $\underline{a}'$.

We show that X^τ_γ is irreducible if and only if $\det R^\gamma_\tau(\lambda) \neq 0$ for all γ in Γ, and that $U(\underline{g})x^\tau_\lambda = X^\tau_\lambda$ if and only if $\det P^\gamma_\tau(\lambda) \neq 0$ for all γ. Here, x^τ_λ is the vector on which K behaves according to τ. It is also shown that $R_\tau^\gamma(\lambda)$ is Hermitian for certain λ in $\underline{a}'_C$, and if $R_\tau^\gamma(\lambda)$ is positive definite for all γ in Γ, X_λ^τ extends to an irreducible unitary representation of G.

Next begins the process of evaluating the matrices P_τ^γ. It is shown in ICSR that P_τ^γ is a one by one matrix if rank G = 1. Corresponding to each positive root $\emptyset$ of $(\underline{g},\underline{a})$ which is not twice a root, there exists a rank one reductive subgroup $G_\emptyset$ of G with Lie algebra $\underline{g}_\emptyset$ = the span of all root spaces for roots which are multiples of $\emptyset$. We show that det P_τ^γ is a product of translates of corresponding P_τ^γ matrices for the subgroups $G_\emptyset$. Thus we are reduced to finding P_τ^γ for a rank one group.

In ICSR, a 2-integer parameterization $\{\gamma_{ij}\}$ of Γ is found in the rank one case, along with elements u_{ij} in H^*, i, j in Z^+, such that $P_0^{\gamma_{ij}} = p_{u_{ij}}^0$. We show that we also have $P_\tau^{\gamma_{ij}} = p_{u_{ij}}^\tau$. To compute the element $p_{u_{ij}}^\tau$, we make use of the projection $u \to \emptyset_u$: $U \to U(\underline{z})U(\underline{a})$, where $\underline{z}$ is the center of $\underline{k}$, $\underline{k}'$ its commutator, from the decomposition $U = (\underline{k}'U + U\underline{n}) \oplus U(\underline{z})U(\underline{a})$. One easily sees that $p_{\emptyset_u}^\tau = p_u^\tau$ for all u in U. Now, using the diagonal map D: $U \to U \otimes U$, the unique algebra homorphism sending X in $\underline{g}$ to $X \otimes 1 + 1 \otimes X$, we find that the projection $\emptyset$ behaves well with respect to D (as does p^0), while p^τ does not. Next we find some information about the $\underline{k}$-submodules of an irreducible finite-dimensional $\underline{g}$-module, for $\underline{g}$ = su(1,1) or su(2,1). This allows us to find some zeros of $p_{u_{ij}}^\tau$ for these algebras. We then reduce the problem for su(n,1) $(n \geqslant 3)$ to that of su(2,1), and use the knowledge of $p_{u_{ij}}^0$ from ICSR to obtain all the $p_{u_{ij}}^\tau$.

Finally, we piece together all these results to find the zeros of det P_τ^γ and det R_τ^γ for a group of arbitrary rank. In the rank one case, $R_\tau^\gamma(\lambda)$ is a 1x1 real matrix which is positive definite if and only if it is positive. We can find exactly those λ for which $R_\tau^\gamma(\lambda) > 0$ for all γ in Γ, and thus find exactly which irreducible principal series representations with a K-stabilized one-dimensional subspace are unitary.

In the case of arbitrary rank, a "critical strip" is found in $\underline{a}'_C$, which, if non-empty, contains $-\rho$. Det $R_\tau^\gamma(\lambda)$ is always positive in this strip if λ satisfies a certain additional condition, in which case $R_\tau^\gamma(\lambda)$ is in fact positive definite, for all γ in Γ.

In section 1, we give the most important results, along with the basic definitions. Section 2 is a more detailed outline of the paper, without proofs.

In chapter II, we derive the relation of R_τ^γ and P_τ^γ to irreducibility, unitarity, and cyclicity. In chapter III, we reduce the calculation of P_τ^γ to the rank one case and evaluate it in that case. In chapter IV, the results are found and applied.

NON-SPHERICAL PRINCIPAL SERIES REPRESENTATIONS OF A SEMISIMPLE LIE GROUP*†

Chapter I

Section 1. Definitions and Major Results.

For the rest of this paper we use a slightly different notation than we used in the introduction. Let G_0 be a connected real reductive non-compact semisimple Lie group with Lie algebra $\underline{g}_R$. We write $\underline{g}$ for the complexification $\underline{g}_R + i\underline{g}_R$. If $\underline{h}$ is any complex subspace of $\underline{g}$, write $\underline{h}_R$ for $h \cap \underline{g}_R$. Now, there exists a unique simply connected complex Lie group with Lie algebra ad $\underline{g}$. Then G acts on $\underline{g}$ and $\underline{g}_R$ by the adjoint representation Ad.

For much of this paper we will deal with the group G and the Lie algebra $\underline{g}$. In chapters II and IV we deal with G_0, and then we assume that G_0 is semisimple; that is, that $\underline{g}_R$ equals its own commutator subalgebra. Recall that a reductive Lie algebra is the sum of its center and its commutator, which is a semisimple Lie algebra. Suppose H is a complex subgroup of G. Then let H_R be the set of elements in H which stabilize $\underline{g}_R$ under Ad.

Now suppose $\underline{g}$ is semisimple. Then there exists an involution θ on $\underline{g}$, commuting with complex conjugation, called the Cartan involution (the facts stated in this section about semisimple Lie groups, without proof, come from 9c., chapters 3, 4, 7, and 8). Let $\underline{k}$ be the +1 eigenspace for θ on $\underline{g}$, $\underline{p}$ the -1 eigenspace. Then $\underline{k}$ is a subalgebra of $\underline{g}$, and if K is the connected subgroup of G with Lie algebra $\underline{k}$, K_R (which has Lie algebra $\underline{k}_R$) is a maximal compact subgroup of G_R.

Further, if we choose an arbitrary maximal abelian subalgebra $\underline{a}$ of $\underline{p}$ ($\underline{p}$ is not a Lie algebra), then the elements of ad $\underline{a}$ are simultaneously diagonalizable, and elements in $\underline{a}_R$ all have real eigenvalues. Let $\underline{a}'$ be the dual of $\underline{a}$, and $\underline{a}'_R$ the real dual of $\underline{a}_R$.

Let $\Delta = \{\emptyset \text{ in } \underline{a}', \emptyset = 0, \mid \text{there exists } 0 \neq X \text{ in } \underline{g}, [H,X] = \langle\emptyset,H\rangle X \text{ for all } H \text{ in } \underline{a}\}$. We call Δ the set of roots of $(\underline{g},a)$. If $r = \dim \underline{a}$, it is possible to find a subset Σ of Δ with r elements $\sigma_1,\ldots,\sigma_r$ such that if $\emptyset$ is in Δ, $\emptyset = \sum_{j=1}^{r} a_j\sigma_j$, where either all the a_j are non-negative or they are all non-positive integers. Such a set Σ is called a set of simple roots, and it is not unique.

Now one knows that if $\emptyset$ is a root, $-\emptyset$ is a root. Further, if $\emptyset$ and $n\emptyset$ are roots, $n = \pm\frac{1}{2}, \pm 1$, or ± 2. If $\emptyset$ is a root, let $\underline{g}^{\emptyset} = \{X \text{ in } \underline{g} \mid [H,X] = \langle\emptyset,H\rangle X \text{ for all } H \text{ in } \underline{a}\}$. Now pick a set of simple roots Σ. Call a root $\emptyset$

*Received by the editors October 3, 1977.

†Research partially supported by the Massachusetts Institute of Technology.

positive if all the a_j defined above are non-negative, and negative if the a_j are all non-positive. Let Δ_+ be the set of positive roots, Δ_- the set of negative roots. Let $\underline{n} = \sum_{\phi \in \Delta_+} \underline{g}^{\phi}$, $\overline{\underline{n}} = \sum_{\phi \in \Delta_-} \underline{g}^{\phi}$.

Then we have the Iwasawa decomposition, which states that $\underline{g} = \underline{k} + \underline{a} + \underline{n}$, $\underline{g}_R = \underline{k}_R + \underline{a}_R + \underline{n}_R$ (direct sum), $G = KAN$, and $G_R = K_R A_R N_R$ (uniquely: if $kan = k'a'n'$, $k = k'$, $a = a'$, $n = n'$). Here, K, A, and N are the connected subgroups of G with Lie algebras $\underline{k}$, $\underline{a}$, and $\underline{n}$. Also, let M be the centralizer of $\underline{a}$ in K (under Ad), with Lie algebra $\underline{m}$. Then $\underline{g} = \overline{\underline{n}} + \underline{m} + \underline{a} + \underline{n}$ as a direct sum, and $\overline{N}MAN$ is an open dense subgroup of G, where $\overline{N}$ has Lie algebra $\overline{\underline{n}}$. This is called the Bruhat decomposition.

The Iwasawa decomposition can be defined for a reductive non-compact Lie group too. Let θ be the Cartan involution on $[\underline{g},\underline{g}]$, and let $\underline{p}$ = (+1 eigenspace of θ) $\oplus$ (center of $\underline{g}$). Than $\underline{a}$ can be chosen so that it contains the center of $\underline{g}$, and we can again write $\underline{g} = \underline{k} + \underline{a} + \underline{n} = \overline{\underline{n}} + \underline{m} + \underline{a} + \underline{n}$, where $\underline{k}$, $\overline{\underline{n}}$, $\underline{m}$, and $\underline{n}$ lie in the commutator of $\underline{g}$. Obviously, all roots vanish on the center of $\underline{g}$.

Also, if G_0 is a real connected non-compact semi-simple Lie group with Lie algebra $\underline{g}_R$, and K_0, A_0, and N_0 are the connected subgroups of G_0 with Lie algebras $\underline{k}_R$, $\underline{a}_R$, and $\underline{n}_R$, then $G_0 = K_0 A_0 N_0$, and the Bruhat decomposition also holds.

Now, if H is a Lie group, let $\hat{H}$ be the set of equivalence classes of finite dimensional irreducible unitary representations of H. We will not always distinguish between a representation and its class. If γ is in $\hat{H}$, we write V_γ for the vector space on which H acts according to γ.

Let M' be the normalizer of $\underline{a}$ in K, and let $W = M'/M$. Then W acts on $\underline{a}$. Further, W acts on $\underline{a}'$ by $\langle w\lambda, H\rangle = \langle \lambda, w^{-1}H\rangle$ for w in W, λ in $\underline{a}'$, H in $\underline{a}$. Also, it is known that there exists a unique element κ in W such that $\kappa(\Delta_+) = \Delta_-$. If λ is in $\underline{a}'$, let $\lambda' = -\kappa\lambda$.

Now, it is known that the exponential map exp: $\underline{a} \to A$ is a Lie group isomorphism, considering $\underline{a}$ as a Lie group under addition. If a is in A, λ in $\underline{a}'$, let $a^\lambda = e^{\langle \lambda, H\rangle}$, where H in $\underline{a}$ is the unique element such that exp (H) = a.

We can now begin to define the principal series of representations of G_0. Let β be in $\hat{M}_0$, λ in $\underline{a}'$. Let V_λ^β be the set of smooth vector-valued functions $\{f\colon G_0 \to V_\beta$ such that $f(gman) = a^{\lambda'}\beta(m^{-1})f(g)$, for all g in G_0, m in M_0, a in A_0, and n in $N_0\}$. We complete V_λ^β to a Hilbert space H_λ^β by the inner product $(f,g) = \int_{K_0/M_0} \langle f(k), \overline{g(k)}\rangle\, dk$, where $\langle\ ,\rangle$ is the inner product on V_β and $\overline{}$ denotes complex conjugation. Since β is unitary, $\langle f(km), \overline{g(km)}\rangle = \langle f(k), \overline{g(k)}\rangle$ for all k in K_0, and m in M_0. Further, even if K_0 is not compact, K_0/M_0 is always compact. Thus, the inner product is well defined.

Now we define a representation of G_0 on the Hilbert space H_λ^β. If f is in H^β_λ, g in G_0, let $(\pi^\beta_\lambda(g)f)(x) = f(g^{-1}x)$ for all x in G_0. The set $\{\pi^\beta_\lambda$, β in $\hat{M}_0$, λ in $\underline{a}'\}$ is called the principal series of representations of G_0. The following facts are known about the principal series (see, e.g., 9c., chapter 8):

1. If H is in $\underline{a}$, let $\rho(H) = \frac{1}{2}\mathrm{tr}\ \mathrm{ad}\ H|_{\underline{n}}$. Then $\pi^\beta_{-\rho+i\nu}$ is unitary if ν is in $\underline{a}'_R$.
2. Let X^β_λ = {f in H^β_λ such that $K_0 f$ spans a finite dimensional subspace of H} (this is the set of K_0-finite vectors). Then X^β_λ is a completely reducible K_0-module, and is dense in H^β_λ. If λ is in $\hat{K}_0$, the multiplicity of γ in X^β_λ (as a K_0-module) equals the multiplicity of β in V_γ as an M_0-module (M_0 is the centralizer of $\underline{a}$ in K_0 under Ad). This is the Frobenius reciprocity therem (see, e.g., 9c., p. 118).
3. As a corollary, H^β_λ has a K_0-fixed vector if and only if β is the identity. Kostant studies these representations in detail in ICSR. Further, H^β_λ has a 1-dimensional subspace on which K_0 acts according to a character τ if and only if $\beta = \tau|_{M_0}$. This is the case we will study.
4. Recall W = M'/M. W also equals M_0'/M_0, where M_0' is the normalizer of $\underline{a}$ in K_0. W acts on $\hat{M}_0$ in the following manner: M_0' normalizes M_0. Thus for w in W, choose a representative m^* in M_0' for w. Let β be in $\hat{M}_0$, m in M_0. Then $(w\beta)(m) = \beta(m^{*-1}mm^*)$. Let ν,λ be in $\underline{a}'_R$, μ,β in $\hat{M}_0$. Then $\pi^\mu_{-\rho+i\nu} = \pi^\beta_{-\rho+i\lambda}$ if and only if there exists w in W such that $w\mu = \beta$, and $w\nu = \lambda$.
5. Suppose $G_0 = G_R$. Then H^β_λ extends to a holomorphic representation of G, and the K-decomposition of X^β_λ is the same as the K_R-decomposition (see ICSR, section 7).
6. X^β_λ is a module for $U = U(\underline{g})$. The action is defined as follows: if f is in X^β_λ, X in $\underline{g}$, g in G_0, then $Xf(g) = d/dt\ (t=0)\ f(\exp(-tX)g)$, where it can be shown that this limit is well defined. Until the last two sections, we will only be concerned about the action of U and K_0 on X^τ_λ, where τ is a character on K_0.

If $\underline{g}$ is reductive, define rank $\underline{g}$ = dim $[\underline{g},\underline{g}]\cap\underline{a}$. Let $\Delta_+^1 = \{\emptyset$ in Δ_+ such that $\frac{1}{2}\emptyset$ is not a root$\}$. If $\emptyset$ is in Δ_+^1, let $\underline{g}_\emptyset = \sum_{j=-2}^{+2} \underline{g}^{j\emptyset}$. Note that $\underline{a} + \underline{m} \subset \underline{g}_\emptyset$, and that $\underline{g}_\emptyset$ is a rank one reductive subalgebra of $\underline{g}$. Let $w_\emptyset$ in $[\underline{g}_\emptyset,\underline{g}_\emptyset]\cap\underline{a}$ be such that $\langle\emptyset,w_\emptyset\rangle = 1$. Let $s_\emptyset = \dim \underline{g}^{2\emptyset}$, let $n_\emptyset = 1$ if $s_\emptyset = 0,2$ if $s_\emptyset>0$; $1_\emptyset = \frac{1}{2}\dim \underline{g}^\emptyset + 1$ if $s_\emptyset>0$, $\frac{1}{2}\dim \underline{g}^\emptyset$ if $s_\emptyset = 0$. Soon we will be able to state main results of this paper. Let τ be a character on K_0. We write τ also for its restriction to M_0, and for its differential on $\underline{k}$. Thus, letting $\underline{k}_\emptyset = \underline{k}\cap\underline{g}_\emptyset$, we see that $\tau|_{\underline{k}_\emptyset}$ is a character on $\underline{k}_\emptyset$. But it is a well known result of E. Cartan (see, e.g., 2a., chapter 9), that if rank $\underline{g}_\emptyset = 1$, then $\underline{k}_\emptyset$ is either semisimple (in which case $\tau|_{\underline{k}_\emptyset} = 0$) or $\underline{g}_\emptyset = su(n,1)$, in which case $\underline{k}_\emptyset$ has a one dimensional center. In this case, the characters of $K_\emptyset \subset G$

are parameterized by the integers, and thus up to sign there exists a smallest character $\tau'_{\emptyset,0}$ on $K_\emptyset$. We also write $\tau'_{0,0}$ for its differential on $\underline{k}_\emptyset$. Thus, $\tau|_{\underline{k}_\emptyset}$ is some (not necessarily integral) multiple $m'_\emptyset$ of the character $\tau'_{\emptyset,0}$. For the purpose of clarity in various formulas, let $m_\emptyset = m'_\emptyset$ if $\underline{g}_\emptyset = su(1,1)$ and $= (n+1)m'_\emptyset$ if $\underline{g}_\emptyset = su(n,1)$, $n \geq 2$. We now state the main results of this paper.

Theorem 1.1. As a $U(\underline{g})$-module, X^τ_λ is irreducible if and only if for all $\emptyset$ in $\Delta_+{}^1$,

(1) $\langle \lambda+\rho, w_\emptyset \rangle - 1_\emptyset \pm m_\emptyset \notin n_\emptyset Z^+$, and

(2) $-\langle \lambda+\rho, w_\emptyset \rangle - 1_\emptyset \pm m_\emptyset \notin n_\emptyset Z^+$.

Throughout this paper, Z^+ = the non-negative integers. Let x^τ_λ in X^τ_λ be such that the action of $\underline{k}$ on x^τ_λ is τ. We say X^τ_λ is cyclic if $Ux^\tau_\lambda = X^\tau_\lambda$.

Theorem 1.2. X^τ_λ is cyclic if and only if for all $\emptyset$ in $\Delta_+{}^1$, $\langle \lambda+\rho, w_\emptyset \rangle - 1_\emptyset \pm m_\emptyset \notin n_\emptyset Z^+$.

Now, for all $\emptyset$ in $\Delta_+{}^1$, let $J_\emptyset \subset R$ (the reals) be the largest open interval containing 0 such that if $t \in J_\emptyset$, $\pm t - 1_\emptyset \pm m_\emptyset \notin n_\emptyset Z^+$. Note that $J_\emptyset$ is empty if and only if $(\pm m_\emptyset - 1_\emptyset) \in n_\emptyset Z^+$. Let CS (for "critical strip") denote $\{\lambda$ in $\underline{a}'$ such that $\langle \lambda+\rho, w_\emptyset \rangle \subset J_\emptyset$ for all $\emptyset$ in $\Delta_+{}^1$. If λ, ν are in $\underline{a}'$, write $\lambda \sim_W \nu$ if there exists w in W such that $w\lambda = \nu$. Let $\underline{a}^* = \{\lambda$ in $\underline{a}'$ such that $(\lambda+\rho) \sim_W (\overline{\lambda}+\rho)\}$. We have

Theorem 1.3. If λ is in $\underline{a}^* \cap$ CS, π^τ_λ is irreducible and unitary.

Note that, unless CS is empty, $\{-\rho+i\nu,\ \nu$ in $\underline{a}'_R\} \subset \underline{a}^* \cap$ CS.

Theorem 1.4. Let G_0 have Lie algebra su(n,1), $n \geq 1$. Then the irreducible unitary representations of G_0 with a vector on which K_0 acts according to a character τ consists of

(a) $\{\pi^\tau_\lambda,\ \lambda \in$ CS $\cap\ \underline{a}^*\}$,

(b) If CS is empty, the additional representations $\{\pi^\tau_{-\rho+i\nu},\ \nu$ in $\underline{a}'_R,\ \nu \neq 0\}$, and

(c) A finite number of additional representations. If $m = m_\emptyset$, this number is 1 if $n = 1$ or $|M| < n$. If $|m| \geq n \geq 2$, this number is $[(m-n)/2) + 3/2]°$, where []° denotes "greatest integer less than."

When results from ICSR are applied, it will sometimes be necessary to deal with an extension of K. This extension is K_θ = {k in G such that k commutes with θ}. It can be shown (see ICSR) that $K_\theta = KF$, where $F \subset \exp(i\underline{a}_R)$ is a finite group of 2^r elements, where r = rank $\underline{g}$.

Section 2. An Outline.

In this section, we outline the methods by which we prove the theorems of section 1. We assume that $\underline{g}$ is reductive unless it is clear from the context that it is semisimple. We fix a character τ: $\underline{k} \to C$. We want $\tau(\underline{k}_R) \subset iR$, so that τ is the differential of a unitary character on the simply connected covering group of K_0.

Let $U = U(\underline{g})$ be the universal enveloping algebra of $\underline{g}$ (see, e.g., 3a., chapter 5). Then $\underline{g}$ is a subalgebra of U, and so are $\underline{k}$, $\underline{a}$, $\underline{n}$, etc. Let $\underline{K}^{\#}$ be the subalgebra of U given by the span of $\{X - \tau(X) \mid X \text{ in } \underline{k}\}$, $\overline{\underline{k}^{\#}}$ the span of $\{X + \tau(X) \mid X \text{ in } \underline{k}\}$. Then one easily sees that elements of $\overline{\underline{k}^{\#}}$ are complex conjugates of elements in $\underline{k}^{\#}$. For, if $X \in \underline{k}_R$, $\overline{X} = X$, and $\overline{\tau(X)} = -\tau(X)$ (here, and always, $\bar{}$ denotes complex conjugation). Obviously, when we write $\underline{k}^{\#}$, we must have a specific character τ in mind.

Now, the Birkhoff-Witt theorem (see 3a., chapter 5) easily implies that $U = (\underline{k}U + U\underline{n}) \oplus U(\underline{a})$. Defining the degree of an element u in U to be the smallest integer n such that $u = X_{11}\cdots X_{1r_2} + X_{21}\cdots X_{2r_2} + \cdots + X_{k1}\cdots X_{kr_k}$, where X_{ij} is in $\underline{g}$, and each $r_i \leq n$, it is an easy induction on the degree of u to show that $u \in (\underline{k}^{\#}U + U\underline{n}) \oplus U(\underline{a})$. Thus let $u \to p_u^{\tau}$ be the projection $U \to U(\underline{a})$, under this direct sum decomposition.

We now define some notation we will use throughout the paper. If V is a vector space, and a Lie group H and Lie algebra $\underline{h}$ act on V, write $V^H = \{v \text{ in } V \mid hv = v \text{ for all } h \text{ in } H\}$ and $V^{\underline{h}} = \{v \text{ in } V \mid Xv = 0 \text{ for all } X \text{ in } \underline{h}\}$. Now, G, G_0, and U act on U by the respective adjoint representations. Thus if V is a subspace of U, H a subgroup of G or G_0, and $\underline{h}$ a subalgebra of U, then V^H and $V^{\underline{h}}$ are as above, with the action assumed to be the adjoint action.

Recall that the Weyl group W acts on $\underline{a}'$. If $w \in W$, let w' be the automorphism of $\underline{a}'$: $w'\lambda = w(\lambda+\rho)-\rho$ for λ in $\underline{a}'$. Let W_T be the group of automorphisms $\{w' \mid w \text{ in } W\}$. Thus W_T acts on $\underline{a}'$, and thus on $U(\underline{a})$, which is canonically isomorphic to the polynomial algebra on $\underline{a}'$ in the following manner: if λ in $\underline{a}'$, $H_1,\ldots,H_r$ in $\underline{a}$, then $H_1\ldots H_r(\lambda) = \langle\lambda,H_1\rangle\langle\lambda,H_2\rangle\cdots\langle\lambda,H_r\rangle$. Harish-Chandra has proved the following theorem for $\tau = 0$ (see 2a., chapter 10);

<u>Theorem 2.1.</u> $I = U^{\underline{k}} \cap \underline{k}^{\#}U = I^{\underline{k}} \cap U\underline{k}^{\#}$ is a 2-sided ideal in $U^{\underline{k}}$. $0 \to I \xrightarrow{\text{incl}} U^{\underline{k}} \xrightarrow{u \to p_u^{\tau}} U(\underline{a})^{W_T} \to 0$ is an exact sequence.

It follows easily that if this statement holds for $\underline{g}$ semisimple, it holds for $\underline{g}$ reductive. The proof in section 3 assumes the former.

The first statement of this theorem is fairly easy to prove. Also, it is fairly obvious that $I \subset \ker p^{\tau}$.

Now, there exists a symmetrization map β: $s(\underline{g}) \to U(\underline{g})$, where $s(\underline{g})$ is the symmetric algebra on $\underline{g}$. β is the unique linear map sending XY to $\frac{1}{2}(XY + YX)$ for X,Y in $\underline{g}$. It is clear that, if a space V is an H-module, then the symmetric algebra S(V) is an H-module in a canonical fashion.

From the Birkhoff-Witt theorem, one sees that $U = \underline{k}U + \beta(S(\underline{p}))$ (see ICSR, section 1.4), and by induction on the degree of an element one sees that $U = \underline{k}^{\#}U + \beta(S(\underline{p}))$. From this it follows that $u^{\underline{k}} = (\underline{k}^{\#}U \cap U^{\underline{k}}) + \beta(S(\underline{p})^{\underline{k}})$. From the structure of $S(\underline{p})^{\underline{k}} = S(\underline{p})^K$ one shows that $p^{\tau}\circ\beta$ is 1-1 on $S(\underline{p})^{\underline{k}}$, and thus $\ker p^\tau = \underline{k}^{\#}U \cap U^{\underline{k}}$. One also sees that, if p maps $U^{\underline{k}}$ into $U(\underline{a})^{W_T}$, it is onto. That $\operatorname{Im} p^\tau \subset U(\underline{a})^{W_T}$ follows from Fact 4 of section 1 (equivalence of π^τ_λ, π^τ_μ if $\lambda \sim_{W_T} \mu$) and:

Proposition 2.2. Let $\lambda \in \underline{a}'$, $u \in U^{\underline{k}}$, $x \in (X^\tau_\lambda)^{\underline{k}^{\#}}$. Then $ux = p^\tau_u(\lambda)x$.

Next we show:

Proposition 2.3. Let Z be an irreducible U-module such that $Z^{\underline{k}^{\#}} \neq 0$. Then $\dim Z^{\underline{k}^{\#}} = 1$.

The proof is similar to the proof of the corresponding result in ICSR (for $\tau = 0$). The main result needed for the proof is Theorem 2.1.

We then find an explicit construction for all such U-modules, based on the similar construction in ICSR.

Proposition 2.4. Let u be in $U^{\underline{k}}$, λ in $\underline{a}'$. Then there exists a unique irreducible U-module Z_λ with $Z_\lambda^{\underline{k}^{\#}} \neq 0$ such that if u is in $U^{\underline{k}}_\lambda$, x in $Z^{\underline{k}^{\#}}$, $u \cdot x = p^\tau_u(\lambda)x$. Every irreducible U-module with a $\underline{k}^{\#}$-annihilated vector is found this way, and if $\mu, \nu \in \underline{a}'$, $Z_{-\rho+\mu} = Z_{-\rho+\nu}$ if and only if $\mu \sim_W \nu$.

We will write Z^τ_λ for Z_λ whenever there is any doubt as to the character λ to which we refer. Obviously, we want to find λ for which $X^\tau_\lambda = Z^\tau_\lambda$, for then X^τ_λ is irreducible.

Now we have to make some more definitions. On a semisimple Lie algebra $\underline{g}$ there exists a bilinear form, $(X,Y) = \operatorname{tr}(\operatorname{ad} X \operatorname{ad} Y)$ (X,Y in $\underline{g}$), called the Killing form. One knows (see 9c., chapter 7) that it is positive definite on $\underline{p}_R$, negative definite on $\underline{k}_R$. If $\underline{g}$ is reductive, we extend the form from $[\underline{g},\underline{g}]$ to $\underline{g}$ arbitrarily, subject to the restriction that it be positive definite on the center of $\underline{g}_R$. We can use the Killing form to define an action of $S(\underline{p})$ as differential operators on $S(\underline{p})$ in the following manner: If $X \in \underline{p}$, let D_X be the differential operator $D_X(Y) = (X,Y)$ for Y in $\underline{p}$; $D_X(pq) = pD_Xq + qD_Xp$ for p, q in $S(\underline{p})$. Further, if p,q are in $S(\underline{p})$, let $D_{pq} = D_pD_q$. Now, let $J = S(\underline{p})^K$, $J_+ = \{q$ in J such that q has no constant term$\}$. Let $H = \{r$ in $S(\underline{p})$ such that $D_qr = 0$ for all q in $J_+\}$. H is called the set of harmonics in $S = S(\underline{p})$. We write J^* for $\beta(J)$, H^* for $\beta(H)$, and S^* for $\beta(S)$.

In section 1.4 of ICSR one proves

Proposition 2.5. $U = U\underline{k} + S^* = \underline{k}U + S^* = U\underline{k} + J^*H^* = \underline{k}U + J^*H^* = U\underline{k} + H^*J^* = \underline{k}U + H^*J^*$.

By a simple induction on the degree of an element one can prove:

Corollary 2.6. All these equalities hold with $\underline{k}^{\#}$ or $\overline{\underline{k}^{\#}}$ replacing $\underline{k}$.

In (5a.) it is shown that $S = J \otimes H$. From this it is shown in ICSR that $S^* = J^* \otimes H^*$. Also in (5a.) one proves:

Proposition 2.7. H and H^* are K_θ-modules. If $\gamma \in \hat{K}_\theta$, the multiplicity of γ in H or H^* is $\dim V_\gamma^{M_\theta}$, where M_θ is the centralizer of $\underline{a}$ in K_θ.

Remark 2.8. The methods of (5a.) can be revised slightly to show that the above statements still hold with K_θ replaced by K, K_R, or K_0, M_θ by M, M_R, or M_0. In ICSR one need only deal with K_θ. Here, we need to deal with K_0 because τ need not be a character on K_θ or K, but only on K_0. The remark above implies, by Frobenius reciprocity, that as K_0-modules, $H = H^* = X_\lambda^o$, λ in $\underline{a}'$.

Now, let (π, V) be a K_0-module, τ a character on K_0. Then we write $\tau \otimes V$ for the K_0-module with underlying vector space V, and K_0-action $\tau \otimes \pi$: $(\tau \otimes \pi)(k)(v) = \tau(k^{-1})\pi(k)v$ for k in K_0, v in V.

Let $A(\lambda)$ be the set of U-modules X, completely reducible under $\underline{k}$, such that $\dim X^{\underline{k}^\#} = 1$, and if $u \in U^{\underline{k}}$, $x \in X^{\underline{k}^\#}$, $u \cdot x = p_u(\lambda)x$. Let $C(\lambda)$ be the subset of $A(\lambda)$ of cyclic U-modules, i.e., $\{X \text{ in } A(\lambda) \text{ such that } U \cdot X^{\underline{k}^\#} = X\}$. Then one easily sees that X_λ^τ is in $A(\lambda)$ and Z_λ^τ in $C(\lambda)$. We prove:

Theorem 2.9.

(a) Z_λ^τ is a K_0-module.

(b) Let $X \in C(\lambda)$, $x \in X^{\underline{k}^\#}$. Let f: $H^* \to \tau \otimes X$ be defined by $f(u) = u \cdot x$. Then f is a $\underline{k}$-linear surjection, and if $X = Z_\lambda^\tau$, f is K_0-linear. Thus, Z_λ^τ is completely reducible as a K_o-module, and if $1(\gamma)$ is the multiplicity of $\gamma \in \hat{K}_0$ in H, and $m_\lambda(\gamma)$ is the multiplicity of γ in $\tau \otimes Z_\lambda^\tau$, then $m_\lambda(\alpha) \leqslant 1(\gamma)$.

(c) The multiplicity of γ in $\tau \otimes X_\lambda^\tau$ is $1(\gamma)$.

The proof of (a) follows from the original construction of Z_λ^τ. The proofs of (b) and (c) are straightforward.

The next step is to define a certain $U(\underline{a})$-valued matrix whose rank for each λ in $\underline{a}'$ is $m_\lambda(\gamma)$. Let $E_\gamma = \mathrm{Hom}_{K_0}(V_\gamma, H^*)$, $E_\gamma' = \mathrm{Hom}_{K_0}(V_\gamma', H^*)$, where V_γ' is the dual of V_γ. Then $\dim E_\gamma = \dim E_\gamma' = 1(\gamma)$. Let $\{v_1, \ldots, v_r\}$ be a basis of V_γ, $\{v_1', \ldots, v_r'\}$ a dual basis of V_γ'. If σ is in E_γ, ν in E_γ', let $z(\sigma, \nu) = \sum_{i=1}^{r} \nu(v_i')\sigma(v_i)$. Let $\{\sigma_1, \ldots, \sigma_{1(\gamma)}\}$ be a basis of E_γ, $\{\nu_1, \ldots, \nu_{1(\gamma)}\}$ a basis for E_γ'. Write $z(i,j) = z(\sigma_i, \nu_j)$. Then let R_τ^γ be the matrix $(R_\tau^\gamma)_{ij} = p_{z(i,j)}^\tau$.

Theorem 2.11. Rank $R_\tau^\gamma(\lambda) = m_\lambda(\gamma)$. Thus X_λ^τ is irreducible if and only if $\det R_\tau^\gamma(\lambda) \neq 0$.

The proof of this theorem is similar to the proof of the corresponding result in ICSR.

Now, for u in U, let u^t be the unique element in U such that:

(1) if u,v are in U, $(uv)^t = v^t u^t$.

(2) if X in $\underline{g}$, $X^t = -X$.

If u is in U, let $p_\tau^u(\lambda) = p_{u^t}^{-\tau}(\lambda')$. Recall $\lambda' = -\kappa\lambda$, where $\kappa \in W$, $\kappa(\Delta_+) = \Delta_-$. If $\{\sigma_1, \ldots, \sigma_{1(\gamma)}\}$ is a basis of E_γ, $\{v_1, \ldots, v_{1(\gamma)}\}$ is a basis of $V_\gamma^{M_o}$, let $(P_\tau^\gamma)_{ij} = p_\tau^{\sigma_i(v_j)}$. The proof of the following is similar to the proof of Theorem 2.11.

Theorem 2.12. X_λ^τ is cyclic if and only if rank $P_\tau^\gamma = l(\gamma)$.

Now, it turns out that R_τ^γ has uses other than for determining irreducibility. Using Nelson's theorem (see section 14), one can show that Z_λ^τ can be extended to a unitary representation of G_0 if and only if there exists a positive definite U-invariant Hermitian form on Z_λ^τ (see section 6). It is straightforward to show:

Proposition 2.13. Z_λ^τ has a U-invariant Hermitian structure if and only if $(\lambda+\rho)\sim_W (\bar{\lambda}+\rho)$.

Next, using methods of ICSR, one shows

Proposition 2.14. Suppose $(\lambda+\rho)\sim_W (\bar{\lambda}+\rho)$. Then there exists a basis with respect to which $R_\tau^\gamma(\lambda)$ is Hermitian. The Hermitian structure on Z_λ^τ is positive definite if and only if $R_\tau^\gamma(\lambda)$ is positive semi-definite.

Further, the rather awkward matrix R_τ^γ can be made more manageable. Using the real bilinear pairing between $X_\lambda^{-\tau}$ and $X_{-\bar{\lambda}-2\rho}^\tau$, one can show

Theorem 2.15. $R_\tau^\gamma(\lambda) = P_\tau^\gamma(-\bar{\lambda}-2\rho)^* P_\tau^\gamma(\lambda)$, where * denotes conjugate transpose.

Now define a matrix P_γ^τ by $(P_\gamma^\tau)_{ij} = p_{\sigma_i}^\tau(v_j)$. We show in section 13 that det P_γ^τ is a scalar multiple of det P_τ^γ. Chapter III is devoted to evaluating det P_γ^τ.

Let $\Gamma = \{\gamma$ in $\hat{K}_\theta$ such that $l(\gamma) \geq 1\}$, $\Gamma_0 = \{\gamma$ in $\hat{K}_0$ such that γ occurs in $H^*\}$. Then there is a natural correspondence between Γ and Γ_0, since the decomposition of H^* into K_θ-, K-, or K_0-modules is the same. Thus, $p_\gamma^\tau = \det P_\gamma^\tau$ depends only on the Lie algebra $\underline{g}$, and not on the group G_0. So, in chapter III we deal with the groups G and K_θ, and the set Γ. This simplifies matters since many results from ICSR, which are formulated in terms of G, are used.

The first step is a reduction to the rank one case. The main tools are Theorem 2.1 (especially the fact that Im $(p^\tau) \subset U(a)^{W_T}$) and the following result from ICSR:

Theorem 2.16. If rank $\underline{g} = 1$, and γ is in Γ, $l(\gamma) = 1$.

To state the reduction theorem, a few definitions are necessary. Let $\phi \in \Delta_+^1$. Recall $\underline{g}_\phi = \sum_{j=-2}^{2} \underline{g}^{j\phi}$, $\underline{k}_\phi = \underline{k} \cap \underline{g}_\phi$, K_ϕ the connected subgroup of G with Lie algebra $\underline{k}_\phi$. Let W be the K_ϕ-module in V_γ generated by $V_\gamma^{M_\theta}$. Then $W = \sum_{i=1}^{l(\gamma)} V_{\delta_i^\phi}$, as a direct sum of K_ϕ-modules, where $\delta_i^\phi \in \Gamma_\phi = \{\gamma$ in $\hat{K}_\phi$ such that $V_\gamma^{M_\theta} \geq 1\}$. Also, let ρ_ϕ in $\underline{a}'$ be such that $\langle \rho_\phi, H\rangle = \frac{1}{2}\mathrm{tr}\, \mathrm{ad}\, H|_{\underline{n}_\phi}$ for all H in $\underline{a}$. If $q \in U(\underline{a})$, λ, μ in $\underline{a}'$, let $T_\lambda q \in U(\underline{a})$: $T_\lambda q(\mu) = q(\mu-\lambda)$. Let $p_\phi = p_{\delta_1^\phi} \cdots p_{\delta_{l(\gamma)}^\phi}$, and let $p_{(\phi)} = T_{\rho_\phi-\rho}\, p_\phi$. Then we have:

Theorem 2.17. Up to a scalar multiple, $p_\gamma^\tau = \prod_{\phi\in\Delta_+^1} p_{(\phi)}$.

This is proved in (6a.) and in section 8. The proof is similar to the corresponding proof in ICSR.

Remark 2.18. It is easy to see that, as γ ranges through Γ, the corresponding $\delta_i^{\emptyset}$ range through $\Gamma_{\emptyset}$. Thus, $p_\gamma^\tau \neq 0$ for all γ in Γ if and only if $p_{\gamma_\emptyset}^\tau = 0$ for all $\gamma_\emptyset$ in $\Gamma_\emptyset$ and all $\emptyset$ in Δ_+^1. So, we are reduced to looking at rank one reductive Lie algebras.

But the only semisimple rank one Lie algebras $\underline{g}$ where $\underline{k}$ is not semisimple are $\{su(n,1),\ n \geq 1\}$.

In section 9, certain results from ICSR are reviewed. An element z in $i\underline{k}_R$ is found, and letting $\underline{d}$ = +1 eigenspace of z in $\underline{p}$, we see that K^z acts on $T = S(\underline{d})$, and it is shown that there exists a natural 1-1 correspondance between K-submodules of H^* and K^z-submodules of $T^* = \beta(T)$: $\gamma \to \gamma_\emptyset$.

Recall from Theorem 2.16 that in the rank one case σ in E_γ and v in V^{M_θ} are unique up to scalar. In ICSR, a certain element w_γ in $(T^*)_{\gamma_\emptyset}$ is found such that $p^0_{\gamma(v)} = p^0_{w_\gamma}$ up to scalar. We prove:

Proposition 2.19. $p_{\sigma(v)} = p^\tau_{w_\gamma}$.

This follows from the fact that, if u lies in an irreducible M-submodule of U which is not isomorphic to the trivial one, $p_u^\tau = 0$. For, p^τ is M-linear, and $U(\underline{a})$ is M-fixed elementwise.

Also, ICSR gives a 2-parameter structure $\{\gamma_{ij}\}$ of Γ and explicitly describes the elements $w_{\gamma_{ij}}$ (see section 9). There are several preliminary steps involved in calculating $p^\tau_{w_{\gamma_{ij}}}$. The first is to introduce the diagonal map D: $U \to U \otimes U$, the unique algebra homomorphism sending X in $\underline{g}$ to $X \otimes 1 + 1 \otimes X$. The following is proved in ICSR:

Proposition 2.20. Let u be in U, $D(u) = \Sigma u_i' \otimes u_i''$. Let λ,ν be in $\underline{a}'$. Then $p^0(\lambda+\nu) = \Sigma p^0_{u_i'}(\lambda) p^0_{u_i''}(\nu)$.

The same result does not hold for p^τ. Instead, we deal with a different projection. Let $\underline{z}$ be the center of $\underline{k}$ (we need not assume rank $\underline{g} = 1$ at this moment), and $\underline{k}'$ its commutator subalgebra. Then $\underline{g} = \underline{k}' + \underline{z} + \underline{a} + \underline{n}$, and thus $U(\underline{g}) = (\underline{k}'U + U\underline{n}) + U(\underline{z})U(\underline{a})$. Let $u \to \emptyset_u$ be the projection to $U(\underline{z})U(\underline{a})$ under this decomposition. Consider $U(\underline{z})U(\underline{a})$ in the natural manner as the polynomial algebra on the dual $\underline{c}'$ of $\underline{c} = \underline{z} + \underline{a}$. The proof of the following is similar to the proof of Proposition 2.20.

Proposition 2.21. Let u be in U, $D(u) = \Sigma u_i' \otimes u_i''$. Let λ,ν $\underline{c}'$. Then $\emptyset_u(\lambda+\nu) = \Sigma \emptyset_{u_i'}(\lambda)\emptyset_{u_i''}(\nu)$.

Note that $p^\tau_{\emptyset_u} = p^\tau_u$, so that it suffices to evaluate $\emptyset_u$. Proposition 2.21 can be generalized to give an expression for $\emptyset_u(t\nu)$ for t in C, $\nu \in \underline{c}'$. The resulting expression is given in section 10.

To evaluate $\emptyset_u$ for u in T^* and $\underline{g}_R = su(N,1)$, there are two distinct cases: N = 1, and $N \geq 2$. If $N \geq 2$, the elements $w_{\gamma_{ij}}$ lie in $U(\underline{s}) \subset U$, where $\underline{s}_R \subset \underline{g}_R$ is isomorphic to su(2,1). We show that $\emptyset_{w_{\gamma_{ij}}}$ is the same, whether calculated with respect to $\underline{s}$ or with respect to $\underline{g}$.

Let α be the simple root in su(N,1). Let τ_0 be the "smallest" character on $\underline{k}$ as defined in the beginning of section 11. If $g_R = su(1,1)$, $\underline{a}$ is a Cartan subalgebra of $\underline{g}$. If $g_R = su(2,1)$, $\underline{m} + \underline{a}$ is a Cartan subalgebra. We use the following result to obtain information about $\emptyset_u$ for u in T^*.

Proposition 2.22.

(a) Let $g_R = su(1,1)$. Let V be an irreducible finite-dimensional U-module with highest weight $n\alpha$, n in Z^+. Then there exists a vector v in V on which $\underline{k}$ acts according to $m\tau_0$ if and only if $m \in Z$, $|m| \leq n$.

(b) Let $g_R = su(2,1)$. Let V_μ be a finite dimensional irreducible U-module with highest weight $\mu, \mu|_{\underline{a}} = n\alpha$, n in Z^+. Then V_μ has a vector on which $\underline{k}$ acts according to $m\tau_0$ if and only if:

(1) $m \in Z$,

(2) $\mu|_{\underline{m}} = m\tau_0|_{\underline{m}}$,

(3) m + n is even, and

(4) $|m| \leq n$.

The proof is in part a modification of a similar result in ICSR. For (b), a detailed examination of the representation theory of sl(3,C) is involved in showing the sufficiency of conditions (1) through (4). Using methods of ICSR, we obtain:

Corollary 2.23.

(a) Let $g_R = su(1,1)$. Let $u \in T_j^*$, the image under β of the homogeneous elements of degree j in T. Let $m \in Z$, $n \in Z^+$, $|m| \leq n$, $j - m > n$. Then $\emptyset_u\ (m\tau_0 + n\alpha) = 0$.

(b) Let $g_R = su(N,1)$, $N \geq 2$. Let $u \in T_j^*$, $m \in Z$, $n \in Z^+$, $j > n \geq |m|$, m + n even. Then $\emptyset_u\ (m\tau_0 + n\alpha) = 0$.

Proposition 2.21 and Corollary 2.23 allow us to evaluate $\emptyset_{w_{\gamma_{ij}}}$. We will give some details for the case of su(N,1), $N \geq 2$. In ICSR one shows

Proposition 2.24. Let $g_R = su(N,1)$, $N \geq 2$. Then there exists elements g, h in $\underline{d}$ such that, as γ runs through Γ, w_γ runs through $\{(g+ih)^j(g-ih)^k$, j, k in $Z^+\}$.

Here, $i = \sqrt{-1}$. Now, direct computation shows $\emptyset_{(g \pm ih)} = w \pm v$, where v in $\underline{z}$ is such that $\tau_0(v) = 1$. We use the generalization of Proposition 2.21 and induction to show $\emptyset_{(g\pm ih)^j} = (w\pm v)(w\pm v-2)\cdots(w\pm v-2j+2)$. Finally we use a double induction and the generalization of Proposition 2.21 to show:

Theorem 2.25. $\emptyset_{(g+ih)^j(g-ih)^k} =$
$(w-v)(w-v-2)\cdots(w-v-2j+2)(w+v)(w+v-2)\cdots(w+v-2k+2)$.

The corresponding result for su(1,1) is easier to prove. These are the last major results we need. Now it is only a question of piecing together the information we have to obtain the theorems of section 1. For, Theorem 2.11, Theorem 2.12, and Propositions 2.13 and 2.14 reduce questions of irreducibility, cyclicity, and unitarity to a knowledge of $R_\gamma^\tau(\lambda)$. Theorem 2.15

essentially reduces the problem to knowing $P^{\tau}_{\gamma}(\lambda)$. Theorem 2.17 reduces the problem of finding det $P^{\tau}_{\gamma}(\lambda)$ to knowing it in the rank one case. In rank one, P^{τ}_{γ} is known from ICSR if $\tau = 0$, as it must be if $\underline{g}_R \neq su(N,1)$. And then Theorem 2.25 finds P^{τ}_{γ} for su(N,1).

The problem which remains involves unitarity: we only know det $R^{\gamma}_{\tau}(\lambda)$, so we do not know when $R^{\gamma}_{\tau}(\lambda)$ is positive semidefinite. Of course, in the rank one case, R^{γ}_{τ} is a 1x1 matrix, and we do know this. But, in general, the best we can do is use a continuity argument to show that $R^{\gamma}_{\tau}(\lambda)$ is positive definite for λ in $\underline{a}^* \cap CS$ (see Theorem 1.3 for notation). Thus we have a sufficient condition for unitarity.

Note that everything we have done applies to non-linear Lie groups. In (9a.) and (9b.) Wallach uses the results of ICSR to find a sufficient condition for irreducibility of the whole non-spherical principal series of certain linear Lie groups. In section 15, we extend these methods to non-linear groups.

CHAPTER II

Section 3. Preliminaries.

Recall from section 2 the projection p^τ: $U \to U(\underline{a})$, and the algebras $\underline{k}^{\#}$ and $\overline{\underline{k}^{\#}}$. In this section we prove:

Theorem 3.1. $I = \underline{k}^{\#}U \cap U^{\underline{k}} = U\underline{k}^{\#} \cap U^{\underline{k}}$ is a 2-sided ideal in $U^{\underline{k}}$. The following is an exact sequence: $0 \to I \xrightarrow{\text{incl.}} U^{\underline{k}} \xrightarrow{u \to p_u} U(\underline{a})^{W_T} \to 0$.

Recall J, H, J^*, and H^* from section 2, and recall (Corollary 2.6) that $U = \underline{k}^{\#}U + J^*H^*$. Now, $\underline{k}^{\#}U$, J^*, and H^* are all $\underline{k}$-modules, so $U^{\underline{k}} = (\underline{k}^{\#}U)^k + (J^*H^*)^{\underline{k}} = \underline{k}^{\#}U \cap U^{\underline{k}} + J^* =$ (replacing τ by $-\tau$) $\overline{\underline{k}^{\#}}U \cap U^{\underline{k}} + J^*$. Also, $\underline{k}^{\#}U \subset [\underline{k}^{\#},U] + U\underline{k}^{\#}$, and the latter are both $\underline{k}$-modules. Thus, $\underline{k}^{\#}U \cap U^{\underline{k}} \subset [\underline{k}^{\#},U]^{\underline{k}} + U\underline{k}^{\#} \cap U^{\underline{k}}$. But it is easy to see that $[\underline{k}^{\#},U] = [\underline{k},U]$, and it is well known that $[\underline{k},U]$ is the sum of all irreducible $\underline{k}$-submodules of U other than the trivial ones. That is, $[\underline{k},U]^{\underline{k}} = 0$. Thus, $U^{\underline{k}} \cap \underline{k}^{\#}U \subset U^{\underline{k}} \cap U\underline{k}^{\#}$, and so $\underline{k}^{\#}U \cap U^{\underline{k}} = U\underline{k}^{\#} \cap U^{\underline{k}}$.

Let $\lambda \in \underline{a}'$. Consider the U-module X_λ^τ of K_0-finite vectors in H_λ^τ (see section 1). Let 1_λ^τ be the unique (by Frobenius reciprocity) $\underline{k}^{\#}$-annihilated vector in X_λ^τ. We have $1_\lambda^\tau(kan) = \tau(k^{-1})a^{\lambda'}$ (here, $k \in K_0$, $a \in A_0$, $n \in N_0$, $\lambda' = -\kappa\lambda$, κ in W, $\kappa\Delta_+ = \Delta_-$). Now we will prove:

Proposition 3.2 For u in $U^{\underline{k}}$, $u \cdot 1_\lambda^\tau = p_u^\tau(\lambda)1_\lambda^\tau$. Proof: Note that $U^{\underline{k}} \cdot 1_\lambda^\tau \subset C \cdot 1_\lambda^\tau$. We first prove:

Lemma 3.3. Let $\delta \in (X_\lambda^\tau)'$, the dual of X_λ^τ, be defined by $\langle \delta,f \rangle = f(e)$, where e is the identity. Then for H in $\underline{a}$, $H \cdot \delta = \langle \lambda',H \rangle \delta$. Here, the action of U on X_λ^τ is defined by $\langle u \cdot x',x \rangle = \langle x',u^t \cdot x \rangle$, where $u \to u^t$: $U \to U$, is defined in section 2, following Theorem 2.11. Proof of lemma: $\langle H \cdot \delta,f \rangle = \langle \delta,-H \cdot f \rangle = -(H \cdot f)(e) = d/dt\ (t=0)\ f(\exp tH) = d/dt\ (t=0)\ f(e)(\exp tH)^{\lambda'} = d/dt\ (t=0)\ f(e)\ e^{t\langle \lambda',H \rangle} = f(e)\langle \lambda',H \rangle = \langle \langle \lambda',H \rangle \delta,f \rangle$. So, $H \cdot \delta = \langle \lambda',H \rangle \delta$.

Now we prove the proposition. Let $u \in U^{\underline{k}}$. Then $\langle \delta,u \cdot 1_\lambda^\tau \rangle = \langle u^t \cdot \delta,1_\lambda^\tau \rangle = \langle p_{u^t}^{-\tau} \cdot \delta,1_\lambda^\tau \rangle$ (since $u^t - p_{u^t}^{-\tau} \in \overline{\underline{k}^{\#}}U + U\underline{n}$, and $\langle (\overline{\underline{k}^{\#}}U + U\underline{n}) \cdot \delta,1_\lambda^\tau \rangle = 0$; for, $(\overline{\underline{k}^{\#}}U)^t = U\underline{k}^{\#}$, which annihilates 1_λ^τ, and $\underline{n}$ annihilates δ). Thus, $\langle \delta,u \cdot 1_\lambda^\tau \rangle = \langle p_{u^t}^{-\tau} \cdot \gamma,1_\lambda^\tau \rangle =$ (by the previous lemma) $p_{u^t}^{-\tau}(\lambda')\langle \delta,1_\lambda^\tau \rangle = p_{u^t}^{-\tau}(\lambda')$ since $1_\lambda^\tau(e) = 1$. Thus $u \cdot 1_\lambda^\tau = p_{u^t}^{-\tau}(\lambda')1_\lambda^\tau$. Now the proposition follows directly from:

Lemma 3.4. Let $u \in U^{\underline{k}}$. Then $p_{u^t}^{-\tau}(\lambda') = p_u^\tau(\lambda)$. Proof of lemma: It is well known (see, e.g., 9c.) that there exists a real bilinear pairing: $X_\lambda^\tau \times X_{-\lambda-2\rho}^{-\tau} \to C$, defined by $(f,g) = \int_{k_0/M_0} f(x)g(x)dx$, such that $(u \cdot f,g) = (f,u^t \cdot g)$ for u in U. So, $(u \cdot 1_\lambda^\tau,1_{-\lambda-2\rho}^{-\tau}) = (1_\lambda^\tau,u^t \cdot 1_{-\lambda-2\rho}^{-\tau})$ or, $p_{u^t}^{-\tau}(\lambda')(1_\lambda^\tau,1_{-\lambda-2\rho}^{-\tau}) = (1_\lambda^\tau,1_{-\lambda-2\rho}^{-\tau})p_u^\tau(-\lambda'-2\rho)$ (note that $\rho' = \rho$). But from Fact 4 of section 1, if $\lambda \in i\underline{a}_R'$, then as U-modules, $X_{\lambda-\rho}^\tau = X^\tau{}_{\lambda-\rho}$ for all s in W. Now, this implies that for u in $U^{\underline{k}}$, $p_u^\tau(\lambda-\rho) = p_u^\tau(s\lambda-\rho)$ if $\lambda \in i\underline{a}_R'$, since the action of $U^{\underline{k}}$ on $X^\tau_{\lambda-\rho}$ is the same as on $X^\tau_{s\lambda-\rho}$. But a polynomial on $\underline{a}'$ is defined by its values on $i\underline{a}_R'$, so $p_u^\tau(\lambda_0-\rho) = p_u^\tau(s\lambda_0-\rho)$ for all λ_0 in $\underline{a}'$.

Setting $\lambda_0 = -\lambda'-\rho$, $s = \kappa$, we get $p_u^\tau(-\lambda'-2\rho) = p_u(\lambda+\rho-\rho)$ (since $\kappa\lambda' = -\lambda$, and $\kappa\rho = -\rho$). Thus, since $p_{u^t}^{-\tau}(\lambda') = p_u^\tau(-\lambda'-2\rho)$, we have $p_{u^t}^{-\tau}(\lambda') = p_u^\tau(\lambda)$.

Q.E.D.

So, we have $u\cdot 1_\lambda^\tau = p_u{}^\tau(\lambda)1_\lambda^\tau$. Further, $p_u^\tau(\lambda-\rho) = p_u^\tau(s\lambda-\rho)$ for all s in W, u in $U^{\underline{k}}$, and λ in $\underline{a}'$. Thus $p_u^\tau \in U(\underline{a})^{W_T}$. Also, we have seen that $\underline{k}^{\#}U \cap U^{\underline{k}} = U\underline{k}^{\#} \cap U^{\underline{k}}$. So, to prove Theorem 3.1, we need only show

(a) ker (p^τ) in $U^{\underline{k}}$ is exactly $\underline{k}^{\#}U \cap U^{\underline{k}}$, and

(b) Im (p^τ) is all of $U(\underline{a})^{W_T}$. Since $U\underline{k}U \cap U^{\underline{k}} + J^*$, (a) and (b) will follow from:

Lemma 3.5. The map $u \to p_u^\tau$ is a linear isomorphism from J^* onto $U(\underline{a})^{W_T}$.

Proof: Recall Chevalley's Theorem (2a., Theorem 6.10 of chapter 10). It states that, if one looks at $S(\underline{p})$ as the algebra of polynomials on $\underline{p}'$, then the restriction map R: $S(\underline{p}) \to S(\underline{a})$ induces a ring isomorphism: $S(p)^{K_0} \xrightarrow{\text{onto}} S(\underline{a})^W$.

Now, one easily sees that $U(\underline{a})^{W_T}$ is a commutative ring canonically isomorphic to $U(\underline{a})^W$. The isomorphism is canonical because $U(\underline{a})^{W_T}$ has the same number of generators, in the same degrees, as $U(\underline{a})^W$. This isomorphism can be given as follows. Let $\{H_1,\ldots,H_r\}$ be a basis of $\underline{a}$. Then if $q \in U(\underline{a})^W$, q is a polynomial $q(H_1,\ldots,H_r)$. Let $q' = q(H_1 + \langle\rho,H_1\rangle,\ldots H_r + \langle\rho,H_r\rangle)$. Then it is easy to see that $q' \in U(\underline{a})^{W_T}$ and that the map $q \to q'$ is an isomorphism. Letting $q_1,\ldots,q_r$ be homogeneous generators of $U(\underline{a})^W$ of degrees $t_1<\ldots<t_r$ (Harish-Chandra shows (1b) that there are r generators, homogeneous of distinct degrees, and that $U(\underline{a})^W$ is a polynomial ring in these generators), then the corresponding q_i' are generators of $U(\underline{a})^{W_T}$. Further, we see deg $q_i' = t_i$, and deg $(q_i-q_i') < t_i$.

Now, it suffices to show that $p^\tau\circ\beta$ is a vector space isomorphism: $S(\underline{p})^{K_0} \to U(\underline{a})^{W_T}$. But it will suffice to show that there exists a set $\{p_1,\ldots,p_r\}$ of generators of $S(\underline{p})^{K_0}$ of degrees $t_1,\ldots,t_r$ such that $p^\tau\circ\beta(p_i)$ are elements of degree t_i with highest order terms equal to q_i. For then $p^\tau\circ\beta$ sends the ring generated by $p_1,\ldots,p_k \in S^{K_0}$ to the ring generated by $q_1',\ldots,q_k'$ in $U(\underline{a})^{W_T}$ for all $k \leq r$. Then $p^\tau\circ\beta$ would be 1-1, onto, and obviously linear.

Let $\underline{v}$ be the perpendicular complement of $\underline{a}$ in $\underline{p}$ under the Killing form. Considering $\underline{a}'$ as a subspace of $\underline{p}'$, we then have $\langle \underline{a}',\underline{v}\rangle = 0$, and $S(\underline{p}) = S(\underline{a}) \oplus \underline{v}S(\underline{p})$. The restriction map R: $S(\underline{p}) \to S(\underline{a})$ is clearly the projection under this decomposition. Also, by Chevalley's theorem, $R\left(S(\underline{p})^{K_0}\right) \subset S(\underline{a})^W$. The theorem allows us to choose generators $\{p_i\}$ such that $R(p_i) = q_i$; then

$p^\tau \circ \beta \circ R(p_i) = q_i$ since $p^\tau \circ \beta =$ identity on $U(\underline{a}) = s(\underline{a})$. So, we need only show $p^\tau \circ \beta\big(R(p_i) - p_i\big)$ has degree $< t_i$.

We show, in fact, that if $H_0 \in \underline{v}$, $H_1, \ldots, H_j \in \underline{p}$, then $\deg p^\tau \circ \beta(H_0 \ldots H_j) \leq j$. This will suffice, since p_i is a sum of terms in $U(\underline{a})$ and of terms in $\underline{v}S(\underline{p})$, and applying $p^\tau \circ \beta(R - \text{identity})$ to $U(\underline{a})$ gives 0. But now write $H_0 = X_1 + X_2$, X_1 in $\underline{k}$, X_2 in $\underline{n}$ ($\underline{v} \subset \underline{k} + \underline{n}$ since $\underline{a}$ is orthogonal to $\underline{k} + \underline{n}$ under the Killing form). By the Birkhoff-Witt theorem, we write $(X_i H_1 \ldots H_j)$ as a sum of terms of the form $Y_1 \ldots Y_k W_1 \ldots W_m Z_1 \ldots Z_m$, where $Y_i \in \underline{k}$, $W_i \in \underline{a}$, $Z_i \in \underline{n}$, and in each term $m \leq j$. Thus we see that $p^\tau \circ \beta(X_i H_1 \ldots H_j)$ has degree $\leq j$, and so $p^\tau \circ \beta\big(R(p_i) - p_i\big)$ has degree $< t_i$, and so $p^\tau \circ \beta$ is a linear isomorphism: $S(\underline{p})^{K_0} \to U(\underline{a})^{W_T}$. Since β is a linear isomorphism: $S(\underline{p})^{K_0} \to J^*$, we have proved the lemma, and thus also Theorem 3.1.

Q.E.D.

As a direct consequence of Theorem 3.1 we get:

Theorem 3.6. Let Z be an irreducible U-module such that $Z^{\underline{k}^\#} \neq 1$.

Proof: The proof of this is almost identical to that of Lemma 1.3.2. of ICSR. Briefly, first an argument due to Dixmier is used to show that the commuting ring of Z consists of scalars. Using this fact and the Jacobson density theorem, one shows that if V is a finite-dimensional subspace of $Z^{\underline{k}^\#}$, and $\sigma \in$ End V, then there exists u in U such that $u \cdot v = \sigma(v)$ for all v in V. Integrating over K_R, one shows that if $\dim Z^{\underline{k}^\#} > 1$, then the image of $U^{\underline{k}}$ (which stabilizes $Z^{\underline{k}^\#}$) in End $Z^{\underline{k}^\#}$ is non-commutative. But one can see that $Z^{\underline{k}^\#}$ is $U^{\underline{k}}$-irreducible, I acts trivially on $Z^{\underline{k}^\#}$, and thus $U^{\underline{k}}/I$, which is commutative by the previous theorem, acts on $Z^{\underline{k}^\#}$. Thus, from the irreducibility, we get that $\dim Z^{\underline{k}^\#} \leq 1$.

Q.E.D.

Section 4. The Irreducible modules Z^τ_λ.

First we prove:

Theorem 4.1. There exists a 1-1 correspondence $\lambda \to Z^\tau_\lambda$ between W_T-orbits in $\underline{a}'$ and irreducible U-modules Z such that $Z^{\underline{k}} \neq 0$. Z^τ_λ is unique, and if $u \in U^{\underline{k}}$, $z^\tau_\lambda \in (Z^\tau_\lambda)^{\underline{k}^\#}$, $u \cdot z^\tau_\lambda = p^\tau_u(\lambda) z^\tau_\lambda$.

We start with a lemma which is also useful at a later time:

Lemma 4.2. Let $v \in U^{\underline{k}}$, $u \in U$. Then $p^\tau_{vu} = p^\tau_v p^\tau_u$. Proof: First, $p^\tau_{vu} - p^\tau_v(p^\tau_u) = 0$. For, let $u = \sum_i \big(X_i - \tau(X_i)\big)u_i + p_u + \sum_j u_j' Y_j$, where $X_i \in \underline{k}$, $u_i' \in U$, $Y_j \in \underline{n}$. Now, since $v \in U^{\underline{k}}$, v and X_i commute, so $v(u - p_u) = \sum_i \big(X_i - \tau(X_i)\big) v u_i + \sum_j v u_j' Y_j \in \underline{k}^\# U + U\underline{n}$. Thus $p^\tau_{v(u - p^\tau_u)} = 0$.

Now, let $v = \sum_i \big(Z_i - \tau(Z_i)\big)v_i + p^\tau_v + \sum_j (v_j' Y_j')$, where $Z_i \in \underline{k}$, $v_j' \in U$, $Y_j' \in \underline{n}$. Then $v p^\tau_u = \sum_i \big(Z_i - \tau(Z_i)\big) v_i p^\tau_u + p^\tau_v p^\tau_u + \sum_j v_j' Y_j' p^\tau_u$. But since $\underline{a}$ normalizes $\underline{n}$, we can move p^τ_u through Y_j', and thus get $v p^\tau_u - p^\tau_v p^\tau_u \in \underline{k}^\# U + U\underline{n}$, or $p^\tau_{v p^\tau_u} = p^\tau_v p^\tau_u$. Combining with the first result, we get $p^\tau_{vu} = p^\tau_v p^\tau_u$.

Q.E.D.

Now, it is easy to see that $U = U^{\underline{k}} \oplus [\underline{k},U]$. Let $u \to u_0$ be the projection: $U \to U^{\underline{k}}$ under this decomposition. Define $L_\lambda = \{u$ in U such that $p_{(vu)_0}(\lambda) = 0$ for all v in $U\}$. L_λ is a proper left ideal of U. We will show that $Z_\lambda = U/L_\lambda$ is the unique U-module with the desired properties.

<u>Lemma 4.3</u>

(a) $\underline{k}^{\#} \subset L_\lambda$.

(b) For all u in $U^{\underline{k}}$, $u - p_u^\tau(\lambda) \subset L_\lambda$.

(c) L_λ is a maximal left ideal in U. Proof of Lemma:

(a) Let $x \in \underline{k}^{\#}$, $u \in U$. Then $ux \in U\underline{k}^{\#}$. Now, $U\underline{k}^{\#}$ and $U^{\underline{k}}$ are $\underline{k}$-modules. So, $(ux)_0 \in U\underline{k}^{\#} \cap U^{\underline{k}} =$ (by Theorem 3.1) $\underline{k}^{\#}U \cap U^{\underline{k}}$. So, since p^τ vanishes on $\underline{k}^{\#}U$, $p^\tau{}_{(ux)_0} = 0$ and thus $x \in L_\lambda$.

(b) Let $u \in U^{\underline{k}}$, $v \in U$. Let $v = v' + v_0$, where $v' \in [\underline{k},U]$ and $v_0 \in U^{\underline{k}}$. Then $(vu)_0 = (v'u)_0 + (v_0u)_0$. But since $u \in U^{\underline{k}}$, $v' \in [\underline{k},U]$, we have $v'u \in [\underline{k},U]$. Thus $(v'u)_0 = 0$. So, $(vu)_0 = (v_0u)_0 = (v_0u)$ since $v_0u \in U^{\underline{k}}$. By Lemma 4.2, $p^\tau_{v_0u} = p^\tau_{v_0}p^\tau_u$. Thus, $p^\tau_{(vu)_0} = p^\tau_{v_0}p^\tau_u$. On the other hand, $p^\tau_{(v\cdot p_u(\lambda))_0} = p^\tau_{v_0\cdot p_u(\lambda)} = p^\tau_{v_0}p_u(\lambda)$. Thus, $p^\tau_{(v(u-p^\tau_u(\lambda))_0}(\lambda) = p^\tau_{v_0}(\lambda)p^\tau_u(\lambda) - p^\tau_{v_0}(\lambda)p^\tau_u(\lambda) = 0$. So, if $u \in U^{\underline{k}}$, $u - p^\tau_u(\lambda) \in L_\lambda$.

(c) Suppose M is a left ideal of U, $L_\lambda \subsetneqq M \subsetneqq U$. Let $u \in M - L_\lambda$. Since $u \notin L_\lambda$, there exists v in U such that $p^\tau_{(vu)_0}(\lambda) \neq 0$. Now, $vu \in M$, since M is a left ideal. Similarly, $\underline{k}^{\#}M \subset M$. But $M\underline{k}^{\#} \subset U\underline{k}^{\#} \subset L_\lambda$ (see (a)). Thus, $[\underline{k}^{\#},M] \subset M$. But since M is a Lie subalgebra of U, it is easy to see that $[\underline{k}^{\#},M] = [\underline{k},M]$. Thus M is a $\underline{k}$-module. So, $(vu)_0 \in M$ since $vu \in M$. But by (b), $(vu)_0 - p^\tau_{(vu)_0}(\lambda) \in L_\lambda \subset M$. Thus, $0 \neq p^\tau_{(vu)_0}(\lambda)$ is a scalar in M. Thus $M = U$, a contradiction, proving (c).

Q.E.D.

Now we prove Theorem 4.1. By (c), L_λ is maximal, so $Z^\tau_\lambda = U/L_\lambda$ is irreducible. Since $\underline{k}^{\#} \subset L_\lambda$, $\underline{k}^{\#}\cdot(1 + L_\lambda) \subset L_\lambda$, so $(1 + L_\lambda) \in (Z^\tau_\lambda)^{\underline{k}^{\#}}$. By Theorem 3.6, $\dim (Z^\tau_\lambda)^{\underline{k}^{\#}} = 1$, so it is generated by $(1 + L_\lambda)$. Now, for u in $U^{\underline{k}}$, $u - p^\tau_u(\lambda) \in L_\lambda$ by (b). So, $\left(u - p^\tau_u(\lambda)\right)\cdot(1 + L_\lambda) \quad L_\lambda$. Thus $u\cdot(1 + L_\lambda) = p^\tau_u(\lambda)(1 + L_\lambda)$. This proves the existence of Z^τ_λ.

To prove uniqueness, let Z be an irreducible U-module, $0 \neq z \in Z^{\underline{k}^{\#}}$, $u\cdot z = p^\tau_u(\lambda)z$ for all u in $U^{\underline{k}}$. Let $M = \{u$ in U such that $u\cdot z = 0\}$. Then, since Z is irreducible, M is a maximal left ideal in U, and $Z = U/M$. We wish to show $M = L_\lambda$, and it clearly suffices to show $M \subset L_\lambda$. Assume there exists u in $M - L_\lambda$. Then there exists v in M such that $p^\tau_{(vu)_0}(\lambda) \neq 0$. So, $vu \in M$. Also, M is a $\underline{k}$-module: $\underline{k}^{\#}M \subset M$, $M\underline{k}^{\#} \subset U\underline{k}^{\#} \subset M$ since $\underline{k}^{\#}z = 0$, and thus as before, $[\underline{k}^{\#},M] = [\underline{k},M] \subset M$. So, $(vu)_0 \in M$, and if $w \in U^{\underline{k}}$, $\left(w - p^\tau_w(\lambda)\right)\cdot z = 0$. Thus, $(vu)_0 - p^\tau_{(vu)_0}(\lambda) \in M$. This yields the contradiction $0 \neq p^\tau_{(vu)_0}(\lambda) \in M$.

To complete the proof of the theorem, we must show that, if Z is an irreducible U-module, $0 \neq z \in Z^{\underline{k}^{\#}}$, then there exists λ in $\underline{a}'$ such that $u \cdot z = p_u^{\tau}(\lambda)z$ for all u in $U^{\underline{k}}$. If $x \in \underline{k}^{\#}$, $u \in U^{\underline{k}}$, clearly $x \cdot u \cdot z = u \cdot x \cdot z = 0$. Thus $U^{\underline{k}}$ stabilizes $Z^{\underline{k}^{\#}}$. Let $f\colon U^{\underline{k}} \to C$ be the function $u \cdot z = f(u)z$. Then f vanishes on $I = U\underline{k}^{\#}$, and thus induces a homorphism $f'\colon U^{\underline{k}}/I = U(\underline{a})^{W_T} \to C$. But W_T acts as a reductive algebraic group on $U(\underline{a})$, and thus the only such homomorphisms f' are found by evaluation on a W_T-orbit of $\underline{a}'$. Thus there exists λ in $\underline{a}'$ such that $f'(p) = p(\lambda)$ for all p in $U(\underline{a})$, that is, such that $f(u) = p_u^{\tau}(\lambda)$ for all u in $U^{\underline{k}}$.

Q.E.D.

Remark 4.4. In the course of the proof we saw that if M is a proper left ideal of U containing $\underline{k}^{\#}$ and $\{u - p_u^{\tau}(\lambda) \mid u \in U^{\underline{k}}\}$, then $m \subset L_\lambda$.

Now we make a few more definitions. Call a U-module X admissible if

(1) $\dim X^{\underline{k}^{\#}} = 1$, and

(2) X is completely reducible as a $\underline{k}$-module.

Call an admissible module X cyclic if $U \cdot X^{\underline{k}^{\#}} = X$. The proof of the following lemma is identical to the last step of the proof of Theorem 4.1.

Lemma 4.5. Let X be an admissible U-module. Then there exists λ in $\underline{a}'$ such that for all u in $U^{\underline{k}}$, x in $X^{\underline{k}^{\#}}$, $u \cdot x = p_u^{\tau}(\lambda)x$.

Let A be the set of admissible U-modules, C the set of cyclic ones. Let $A(\lambda) = \{X \text{ in } A \text{ such that for all } u \text{ in } U^{\underline{k}},\ u \cdot x = p_u^{\tau}(\lambda)x\}$, $C(\lambda) = \{X \text{ in } C \text{ such that for all } u \text{ in } U^{\underline{k}},\ x \text{ in } X^{\underline{k}^{\#}},\ u \cdot x = p_u^{\tau}(\lambda)x\}$.

Remark 4.6. $Z_\lambda^{\tau} \in C(\lambda)$. Recall the principal series U-module X_λ^{τ}. It is well known that $X_\lambda^{\tau} \in A$, and in section 3 it was shown that $X_\lambda^{\tau} \in A(\lambda)$.

Let $L_\lambda^{\min}$ be the left ideal in U generated by $\underline{k}^{\#} \cup \{u - p_u^{\tau}(\lambda) \mid u \in U^{\underline{k}}\}$.

Remark 4.7. $L_\lambda^{\min} \subset L_\lambda$. Further, if $Y \in C(\lambda)$, and $L = \{u \text{ in } U \mid u \cdot Y^{\underline{k}^{\#}} = 0\}$, then $L_\lambda^{\min} \subset L \subset L_\lambda$.

Let $Y_\lambda = U/L_\lambda^{\min}$.

Lemma 4.8. $K_0 \cdot L_\lambda \subset L_\lambda$ and $K_0 \cdot L_\lambda^{\min} \subset L_\lambda^{\min}$. Thus Y_λ and Z_λ are K_0-modules.

Proof: To show $L_\lambda^{\min}$ is K_0-stable, it suffices to show that the sets $\underline{k}^{\#}$ and $\{u - p_u^{\tau}(\lambda) \mid u \in U^{\underline{k}}\}$ are K_0-stable. Since $\underline{k}^{\#}$ is clearly K_0-stable, we need only show, for $a \in K_0$, $u \in U^{\underline{k}}$, that $p_{a\cdot u}^{\tau} = p_u^{\tau}$. For then, $a \cdot \left(u - p_u^{\tau}(\lambda)\right) = a \cdot u - p_{a \cdot u}^{\tau}(\lambda)$.

But $U^{\underline{k}} = \underline{k}^{\#}U \cap U^{\underline{k}} + J^*$ (see section 3). So, if $u \in U^{\underline{k}}$, $u = u_1 + u_2$, $u_1 \in \underline{k}^{\#}U \cap U^{\underline{k}}$, $u_2 \in J^*$. Since p^{τ} annihilates $\underline{k}^{\#}U$, $p_u^{\tau} = p_{u_2}^{\tau}$. Further, $\underline{k}^{\#}U \cap U^{\underline{k}}$ is a K_0-module, so $a \cdot u_1 \in \underline{k}^{\#}U \cap U^{\underline{k}}$. Now, $u_2 \in J^* = \beta\left(S(\underline{p})^{\underline{k}}\right) = \beta\left(S(\underline{p})^{K_0}\right)$, so $a \cdot u_2 = u_2$. Thus, $p_u^{\tau} = p_u^{\tau} = p_{a\cdot u_2}^{\tau} = p_{(a\cdot u_1 + a\cdot u_2)}^{\tau} = p_{a\cdot u}^{\tau}$. So, we have shown that $L_\lambda^{\min}$ is K_0-stable.

Now, by Remark 4.4, L_λ is the unique maximal left ideal containing $L_\lambda^{\min}$. So, if $a \in I_0$, $a \cdot L_\lambda$ is maximal, and contains $a \cdot L_\lambda^{\min} = L_\lambda^{\min}$. Thus $a \cdot L_\lambda = L_\lambda$, and so K_0 stabilizes L_λ.

Q.E.D.

Now, suppose π: $\underline{k} \to \text{End } V$ is a representation of $\underline{k}$. Then we construct a new representation $\tau \otimes \pi$: $\underline{k} \to \text{End } V$ by $(\tau\otimes\pi)(X)v = (\pi(X) - \tau(X))v$ for X in $\underline{k}$, v in V. It is easy to see that $\tau \otimes \pi$ is in fact a representation of $\underline{k}$. If V is a $\underline{k}$-module, we will write $\tau \otimes V$ for the $\underline{k}$-module with the $\underline{k}$-action described above. Recall that H^* is a $\underline{k}$-module.

Proposition 4.9. Let $X \in C(\lambda)$. Let $x \in X^{\underline{k}^{\#}}$. Define f: $H^* \to \tau \otimes X$ by $f(u) = u\cdot x$. Then f is a $\underline{k}$-linear surjection.

Proof: Let $u \in H^*$, $Y \in \underline{k}$. Then $f([Y,u]) = [Y,u]\cdot x = Y\cdot u\cdot x - u\cdot Y\cdot x = Y\cdot u\cdot x - u\cdot\tau(Y)x = (Y - \tau(Y))\cdot u\cdot x$ = (calling π the action of $\underline{k}$ on X) $(\tau \otimes \pi)(Y)(u\cdot x) = (\tau \otimes \pi)(Y)f(u)$, proving $\underline{k}$-linearity.

To prove surjectivity, note that $U = U\underline{k}^{\#} + H^*J^*$ (Corollary 2.6) and $U\cdot x = X$. Now, $\underline{k}^{\#}\cdot x = 0$, $J^* \subset U^{\underline{k}}$, and $U^{\underline{k}}\cdot x \subset \mathbb{C}x$. So, $U\cdot x = H^*\cdot x = X$. So, $f(H^*) = X$, proving surjectivity.

Q.E.D.

Now, recall from Lemma 4.8 that Z_λ^τ is a K_0-module. Call π the action of K_0 on Z_λ^τ. Then we have a K_0-module $(\tau \otimes \pi, \tau \otimes Z_\lambda^\tau)$ with underlying vector space Z_λ^τ and K_0-action $(\tau \otimes \pi)(a)v = \pi(a)\tau(a)^{-1}v$. For an arbitrary K_0-module V we define a new K_0-module $\tau \otimes V$ in the same manner. It is easy to see that f: $H^* \to \tau \otimes Z_\lambda^\tau$ is K_0-linear as well.

Remark 4.10. Thus, since $\tau \otimes Z_\lambda^\tau$ is completely reducible as a K_0-module, we see that if $\gamma \in \Gamma_0$ (the elements of K_0 such that $V_\gamma^{M_0} \neq 0$), then the multiplicity of γ in $\tau \otimes Z_\lambda^\tau \leq$ the multiplicity of γ in H^* = (by 5a.) $\dim (V_\gamma)^{M_0}$ = (by definition) $l(\gamma)$.

Call $m_\lambda(\gamma)$ the multiplicity of γ in $\tau \otimes Z_\lambda^\tau$. Recall $Y_\lambda = U/L_\lambda^{\min} \in C(\lambda)$ is a K_0-module.

Lemma 4.11. The map f: $H^* \to \tau \otimes Y_\lambda$ of Proposition 4.9 is a K_0-linear bijection. Moreover, Y_λ is the only element in $C(\lambda)$ with this property. Thus, $H^* = \tau \otimes Y_\lambda$ as K_0-modules.

Proof: The proof of this lemma is identical to that of Lemma 1.5.6 of ICSR. We outline the proof here. For $\lambda \in \underline{a}'$, letting $C_\lambda = \{u - p_u^\tau(\lambda) \mid u \in U^{\underline{k}}\}$, one shows $L_\lambda^{\min} = UC_\lambda + U\underline{k}^{\#} = (U\underline{k}^{\#} + H^*J^*)C_\lambda + U\underline{k}^{\#} = H^*C + U\underline{k}^{\#}$. Further, letting $J_\lambda{}^* = \{u - p_u^\tau(\lambda) \mid u \in J^*\}$, one sees $J^* = \mathbb{C}\cdot 1 + J_\lambda{}^*$. Also, $L_\lambda^{\min} = H^*J^* + U\underline{k}^{\#}$. Further, $U = H^* + H^*J_\lambda{}^* + U\underline{k}^{\#}$, and thus $U = H^* + L_\lambda^{\min}$, as a direct sum of $\underline{k}$-modules. Thus $Y_\lambda = H^*$ as a $\underline{k}$-module. It follows easily that Y_λ is completely reducible as a $\underline{k}$-module and cyclic. Further, the fact that f is a bijection follows from $U = H^* + L^{\min}$. The uniqueness of Y_λ also follows easily.

Now, from Remark 4.7, Remark 4.10, and Lemma 4.11, we get

Corollary 4.12. There exists a unique element in $C(\lambda)$ if and only if

$l(\gamma) = m_\lambda(\gamma)$ for all $\gamma \in \Gamma_0$.

Section 5. Irreducibility and Cyclicity.

Recall the principal series modules X_λ^τ. We first prove:

Proposition 5.1. Let $\gamma \in \Gamma_0$. Then the multiplicity of γ in $\tau \otimes X_\lambda^\tau$ is $l(\gamma)$. Proof: Let 1_λ^τ in $(X_\lambda^\tau)^{\underline{k}\#}$ be the function $1_\lambda^\tau(kan) = \tau(k^{-1})a^{\lambda'}$. Recall the spherical principal series U-module X_λ^0, and let T_λ: $X_\lambda^0 \to \tau \otimes X_\lambda^\tau$ be the map: $(g \in G_0)$ $T_\lambda f(g) = 1_\lambda^\tau(g)f(g)$. Clearly $\mathrm{Im}(T_\lambda) \subset \tau \otimes X_\lambda^\tau$. It will suffice to show that T_λ is a K_0-linear bijection, since Frobenius reciprocity gives us that the multiplicity of γ in X_λ^0 is $l(\gamma)$.

To show K_0-linearity, let $k \in K_0$, $f \in X_\lambda^0$. Then $T_\lambda(k\cdot f)(g) = 1_\lambda^\tau(g)(k\cdot f)(g)$ = (setting $g = k_0an$) $a^{\lambda'}\cdot\tau(k_0^{-1})f(k^{-1}k_0an) = \tau(k^{-1})\tau\left((k^{-1}k_0)^{-1}\right)f(k^{-1}k_0an)\cdot a^{\lambda'} = \tau(k^{-1})1_\lambda^\tau(k^{-1}g)f(k^{-1}g) = (\tau \otimes \pi_\lambda^\tau)(k)(1_\lambda^\tau f)(g) = (\tau \otimes \pi_\lambda^\tau)(k)\ (T_\lambda f)(g)$, thus proving K_0-linearity.

To show bijectivity, we will exhibit a K_0-linear map S_λ: $\tau \otimes X_\lambda^\tau \to X_\lambda^0$ such that $T_\lambda \circ S_\lambda$ = id., and $S_\lambda \circ T_\lambda$ = id. Define $S_\lambda f(g) = 1_\lambda^{-\tau}(g)f(g)$. The K_0-linearity of S_λ follows from the same argument as established the K_0-linearity of T_λ. Now, if $f \in X_\lambda^0$, $(S_\lambda T_\lambda f)(g) = S_\lambda(1_\lambda^\tau f)(g)$. Now, $(1_\lambda^\tau f)(kan) = \tau(k^{-1})f(k)a^{\lambda'}$. So, $S_\lambda(1_\lambda^\tau f)(g) = 1_\lambda^{-\tau}(kan)\tau(k^{-1})f(k)a^{\lambda'} = \tau(k^{-1})\tau^{-1}(k^{-1})f\ (k)a^{\lambda'} = f(kan)$. Similarly, $T_\lambda \circ S_\lambda$ = identity.

Q.E.D.

Define $X_\lambda^{\tau,e} = U\cdot 1_\lambda^\tau \subset X_\lambda^\tau$. Call $n_\lambda(\gamma)$ the multiplicity of γ in Γ_0 in $\tau \otimes X_\lambda^{\tau,e}$. Clearly $n_\lambda(\gamma) \leq l(\gamma)$. Further, $X_\lambda^{\tau,e} \in C(\lambda)$.

Corollary 5.2. X_λ^τ is irreducible if and only if $l(\gamma) = m_\lambda(\gamma)$ for all γ in Γ_0.

Proof: If X_λ^τ is irreducible, $X_\lambda^\tau = Z_\lambda^\tau$ (since there exists a unique irreducible element in $C(\lambda)$). Thus, by Proposition 5.1, $m_\lambda(\gamma) = l(\gamma)$. Conversely, if $m_\lambda(\gamma) = l(\gamma)$ for all γ in Γ_0, by Corollary 4.12 there exists a unique element in $C(\lambda)$. Thus, $Z_\lambda^\tau = X_\lambda^{\tau,e}$, but thus $n_\lambda(\gamma) = m_\lambda(\gamma) = l(\gamma)$, and so $\tau \otimes (X^{\tau,e}) = \tau \otimes (X_\lambda^\tau)$, or $X_\lambda^\tau = Z_\lambda^\tau$, which is irreducible.

Q.E.D.

Now let $\{v_1,\ldots,v_r\}$ be a basis of V_γ, and $\{v_1',\ldots,v_r'\}$ a dual basis of V_γ'. Recall $E_\gamma = \mathrm{Hom}_{K_0}(V_\gamma,H^*)$, and $E_\gamma' = \mathrm{Hom}_{K_0'}(V_\gamma',H^*)$. From (5a.) one knows $\dim E_\gamma = \dim E_\gamma' = l(\gamma)$. If $\sigma \in E_\gamma$, $\nu \in E_\gamma'$, let $z(\nu,\sigma) = \sum_{i=1}^{r} \nu(v_i')\sigma(v_i)$. It is easy to see that $z(\nu,\sigma)$ is K_0-invariant and independent of basis. So, $z(\nu,\sigma) \in U^{\underline{k}}$. Now let $\{\sigma_1,\ldots,\sigma_{l(\gamma)}\}$ be a basis of E_γ, $\{\nu_1,\ldots,\nu_{l(\gamma)}\}$ a basis of E_γ'. Let $z(i,j) = z(\nu_i,\sigma_j)$. Recall the matrix R_τ^γ with entries in $U(\underline{a})$: $(R_\tau^\gamma)_{ij} = p_{z(i,j)}^\tau$. If we change bases, we obtain a new matrix $R_\tau^{\gamma'} = AR_\tau^\gamma B$, where A and B are non-singular matrices with scalar entries. Thus, rank $\left(R_\tau^\gamma(\lambda)\right)$ is independent of bases for all λ in $\underline{a}'$.

Since $H^* \subset U([\underline{g},\underline{g}]) \subset U$, we have that, if λ, $\lambda_0 \in \underline{a}'$, $\lambda_0|_{\underline{a}\cap[\underline{g},\underline{g}]} = 0$, then $R_\tau^\gamma(\lambda + \lambda_0) = R_\tau^\gamma(\lambda)$. Thus in our calculations we can assume $\underline{g}$ is semi-simple. We now have:

Theorem 5.3. For all γ in Γ, $\gamma \in \underline{a}'$, $m_\lambda(\gamma) = \operatorname{rank} R_\tau^\gamma(\lambda)$. Proof: The proof of this is identical to the proof of Lemma 1.5.12 in ICSR. We outline the proof here. Let $E_\gamma^0 = \{\sigma \in E_\gamma \mid \sigma(V_\gamma)\cdot Z_\lambda^{\tau\underline{k}^\#} = 0\}$. One first proves:

Lemma 5.4. Let $\sigma \in E_\gamma$. Then $\sigma \in E_\gamma^0$ if and only if for all $\nu \in E_\gamma{}'$, $p_{z(\nu,\sigma)}^\tau(\lambda) = 0$.

It follows from the fact that $z(\nu,\sigma)$ is K_0-fixed that $\sigma \in E_\gamma^0$ if and only if for all u in $\sigma(V_\gamma)$, v in H^*, $p_{(vu)_0}^\tau(\lambda) = 0$, where the map $u \to u_0$ is defined in section 4. Recall the version of Schur's Lemma which states that if γ and δ are in $\hat{K}_0$, $(V_\gamma \otimes V_\delta)^{K_0} \neq 0$ if and only if γ and δ are dual, in which case its dimension is 1. One uses the previous statement and the first part of Schur's Lemma to show $\sigma \in E_\gamma^0$ if and only if for all $\nu \in E_\gamma{}'$, $p_{(vu)_0}^\tau(\lambda) = 0$ for all u in $\sigma(V_\gamma)$, v in $\nu(V_\gamma{}')$. Then the second statement in Schur's Lemma shows that in this case $p_{(vu)_0}^\tau(\lambda)$ is a scalar multiple of $p_{z(\nu,\sigma)}^\tau$, and thus the lemma follows.

Now, Theorem 5.3 is proved from Lemma 5.4 in exactly the same manner that Theorem 5.7 is proved from Corollary 5.9. The latter proof will be done in detail.

Corollary 5.5. Let $r_\tau^\gamma(\lambda) = \det R_\tau^\gamma(\lambda)$. Then $m_\lambda(\gamma) = 1(\gamma)$ if and only if $r_\tau^\gamma(\lambda) \neq 0$. Thus X_λ^τ is irreducible if and only if $r_\tau^\gamma(\lambda) \neq 0$ for all γ in Γ_0.

Let $u \in U$. Recall $p_\tau^u \in U(\underline{a})$ is such that $p_\tau^u(\lambda) = p_{u^t}^{-\tau}(\lambda')$ for all λ in $\underline{a}'$.

Remark 5.6. For u in $U^{\underline{k}}$, $p_\tau^u = p_u^\tau$ by Lemma 3.4.

Let $\{\sigma_1,\ldots,\sigma_{1(\gamma)}\}$ again be a basis for E_γ, $\{v_1,\ldots,v_{1(\gamma)}\}$ a basis for $(V_\gamma)^{M_0}$. Recall the $U(\underline{a})$-valued matrix P_τ^γ with entries $(p_\tau^\gamma)_{ij} = p^{\sigma_i(v_j)}$. Again, rank P_τ^γ is basis-invariant. We have:

Theorem 5.7. $n_\lambda(\gamma) = \operatorname{rank} P_\tau^\gamma(\lambda)$ for all γ in Γ_0, λ in $\underline{a}'$.

Proof: We first prove:

Lemma 5.8. Let $L = \{u \text{ in } U \text{ such that } u\cdot 1_\lambda^\tau = 0\}$. Then $u \in L$ if and only if, for all a in K_0, $p_\tau^{a\cdot u}(\lambda) = 0$. Thus L is stable under K_0. Proof: For u in U, $K_0\cdot u = K\cdot U = U(\underline{k})u$, so it suffices to prove $u \in L$ if and only if $p^{U(\underline{k})u}(\lambda) = 0$. Recall $\delta \in (X_\lambda^\tau)'$, $\langle\delta,f\rangle = f(e)$. Recall $\underline{n}\cdot\delta = 0$. Note that for all v in U, $v^t - p_{vt}^{-\tau} \in \underline{k}^\# U + U\underline{n}$. Thus, $v - (p_{vt}^{-\tau})^t \in U\underline{k}^\# + n\cdot U$. Thus, $\langle\delta,v\cdot 1_\lambda^\tau\rangle = \langle\delta,(p_{vt}^{-\tau})^t\cdot 1_\lambda^\tau\rangle = \langle p_{vt}^{-\tau}\cdot\delta,1_\lambda^\tau\rangle = p_{vt}^{-\tau}(\lambda')\langle\delta,1_\lambda^\tau\rangle$ by Lemma 3.3 $= p_\tau^v(\lambda)$. Thus, we get that $p_\tau^{U(\underline{k})\cdot u}(\lambda) = 0$ if and only if $\langle\delta,U(\underline{k})\cdot u\cdot 1_\lambda^\tau\rangle = 0$ if and only if $\langle U(\underline{k})\cdot\delta,u\cdot 1_\lambda^\tau\rangle = 0$ if and only if $\langle U(\underline{k})U(\underline{a})U(\underline{n})\cdot\delta,\ u\cdot 1_\lambda^\tau\rangle = 0$ (since $U(\underline{a})$ $U(\underline{n})\cdot\delta = C\delta$) if and only if $\langle U\cdot\delta,u\cdot 1_\lambda^\tau\rangle = 0$ if and only if $u\cdot 1_\lambda^\tau = 0$ if and only if (by definition of L) $u \in L$. This proves the lemma.

Corollary 5.9. Let $\sigma \in E_\gamma$. Then $\sigma(V_\gamma)\cdot 1_\lambda^\tau = 0$ if and only if $p_\tau^{\sigma(v_i)}(\lambda) =$

0 for $i = 1,\ldots,l(\gamma)$. Proof: Let $v \in V_\gamma$. Then $\sigma(v)\cdot 1_\lambda^\tau = 0$ if and only if $p_\tau^{K_0\cdot\sigma(v)}(\lambda) = 0$ if and only if (since σ is K_0-linear) $p_\tau^{\sigma(K_0\cdot v)}(\lambda) = 0$ if and only if $p^{\sigma(V_\gamma)}(\lambda) = 0$. But $V_\gamma = (V_\gamma)^{M_0} + V'$, where V' is the sum of the non-trivial primary M_0-submodules of V_γ. But $p^{\sigma(V')} = 0$ because p_τ is M_0-linear and $U(\underline{a})$ is M_0-fixed. Thus $p_\tau^{\sigma(V_\gamma)}(\lambda) = 0$ if and only if $p^{\sigma(v_i)}(\lambda) = 0$ for $i = 1,\ldots,l(\gamma)$.

Q.E.D.

Now we prove Theorem 5.7. Let $E_\gamma^0 = \{\sigma \in E_\gamma$ such that $\sigma(V_\gamma)\cdot 1_\lambda^\tau = 0\}$. Let $k = \dim E_\gamma^0$. We change our basis $\{\sigma_i\}$ of E_γ so that $\{\sigma_1,\ldots,\sigma_k\}$ is a basis of E_γ^0. Now $p_\tau^{\sigma_i(v_j)}(\lambda) = 0$ by Corollary 5.9 if $i \leq k$ and j is arbitrary. So, $P_\tau^\gamma(\lambda)$ has rank $\leq l(\gamma) - k$. We show rank $P_\tau^\gamma(\lambda) = l(\gamma) - k$ by showing that the following $\big(l(\gamma) - k\big)$ row vectors of $P_\tau^\gamma(\lambda)$ are linearly independent. The vectors are $c_i = \big(p_\tau^{\sigma_i(v_1)}(\lambda),\ldots,p_\tau^{\sigma_i(v_{l(\gamma)})}(\lambda)\big)$ where $(k+1) \leq i \leq l(\gamma)$. Suppose that we have $\sum_{i=k+1}^{\ell(\gamma)} t_i c_i = 0$. Then $0 = \sum_{i=k+1}^{\ell(\gamma)} t_i p_\tau^{\sigma_i(v_j)}(\lambda)$ for each j, $1 \leq j \leq l(\gamma)$. But defining $\sigma_0 = \sum_{i=1}^{\ell(\gamma)} t_i\sigma_i$, we have $p_\tau^{\sigma_0(v_j)}(\lambda) = 0$ for all j. Thus $\sigma_0 \in E_\gamma^0$. But σ_0 is a linear combination of $\sigma_{k+1},\ldots,\sigma_{l(\gamma)}$, and so $\sigma_0 = 0$. So, all the $t_i = 0$, and thus the c_i are linearly independent. So, rank $P_\tau^\gamma(\lambda) = l(\gamma) - k$. So, to prove the theorem, we need only show $n_\lambda(\gamma) = l(\gamma) - k$.

But recall $f\colon H^* \to \tau \otimes (X_\lambda^{\tau,e})$ is a K_0-module surjection by Proposition 4.9, where $f(u) = u\cdot 1_\lambda^\tau$. For any K_0-module W, let W_γ = the sum of the γ-primary submodules of W. Then f clearly maps $(H^*)_\gamma$ onto $(\tau \otimes X_\lambda^{\tau,e})_\gamma$. Further, $(H^*)_\gamma = \sigma_1(V_\gamma) + \ldots + \sigma_{l(\gamma)}(V_\gamma)$, while $(\tau \otimes X_\lambda^{\tau,e})$ contains $n_\lambda(\gamma)$ copies of V_γ. But $f\big(E_\gamma^0(V_\gamma)\big) = 0$, that is, $f\big(\sigma_1(V_\gamma)\big) = 0 = \ldots = f\big(\sigma_k(V_\gamma)\big)$, while f is a linear isomorphism on the subspace of H^* spanned by $\sigma_{k+1}(V_\gamma),\ldots,{}_{l(\gamma)}(V_\gamma)$. Thus, $\{f\big(\sigma_{k+1}(V_\gamma)\big),\ldots,f\big(\sigma_{l(\gamma)}(V_\gamma)\big)\}$ are linearly independent subspaces of $(\tau \otimes X_\lambda^{\tau,e})_\gamma$. Thus there are $\big(l(\gamma) - k\big)$ copies of V_γ in $(\tau \otimes X_\lambda^{\tau,e})$. That is, $n_\lambda(\gamma) = l(\gamma) - k$, proving the theorem.

Q.E.D.

Corollary 5.10. X_λ^τ is cyclic if and only if $p_\tau^\gamma(\lambda) = \det P_\tau^\gamma(\gamma) \neq 0$ for all $\gamma \in \Gamma_0$. Proof: Then, for all γ in Γ_0, rank $P_\tau^\gamma(\lambda) = l(\gamma) = n_\lambda(\gamma)$, and thus $X_\lambda^\tau = X_\lambda^{\tau,e}$. Conversely, we see that if $X_\lambda^\tau = X_\lambda^{\tau,e}$, $n_\lambda(\gamma) = l(\gamma)$ for all γ in Γ_0.

Section 6. Unitarity.

If $u \in U$, let $u^s = \bar{u}^t$. Let $\langle\,,\rangle$ be a Hermitian bilinear form of Z_λ^τ. That is, if $v, v', v'' \in Z_\lambda^\tau$, $\alpha \in C$, then $\langle v' + v'', v\rangle = \langle v', v\rangle + \langle v'', v\rangle$, $\langle \alpha v', v\rangle = \alpha\langle v', v\rangle$, and $\langle v, v'\rangle = \overline{\langle v', v\rangle}$. We call $\langle\,,\rangle$ U-invariant if for all u in U, $\langle u\cdot v, v'\rangle = \langle v, u^s\cdot v'\rangle$. We call the form positive definite if $\langle v,v\rangle > 0$ if $v \neq 0$. Let $\lambda \in \underline{a}'$, $\lambda = -\rho+\mu$. Let $z_\lambda^\tau \in (Z_\lambda^\tau)^{\underline{k}^\#}$. We will prove:

Theorem 6.1. There exists a non-trivial U-invariant Hermitian form on Z_λ^τ if and only if $\mu \sim_W -\bar{\mu}$. If such a form exists, it is unique up to a real scalar and can be chosen so that $\langle z_\lambda^\tau, z_\lambda^\tau\rangle = 1$, and $\langle u\cdot z_\lambda^\tau, v\cdot Z_\lambda^\tau\rangle = p^\tau_{(v^s u)_0}(\lambda)$ for all u, v in U. Proof: Suppose $\langle,\rangle$ exists. Let $W = \{x \in Z_\lambda^\tau$ such that $\langle x,y\rangle = 0$ for all y in $Z_\lambda^\tau\}$. Then W is clearly U-invariant. Since $\langle,\rangle$ is non-trivial, $W \neq Z_\lambda^\tau$, so by irreducibility, W = 0. Thus $\langle,\rangle$ is non-singular.

In fact, $\langle,\rangle$ is non-singular on $(Z_\lambda^\tau)^{\underline{k}^\#}$. For, since Z_λ^τ is completely reducible as a $\underline{k}$-module, let V = the sum of the primary $\underline{k}$-submodules of Z_λ^τ not of type τ. Thus, $Z_\lambda^\tau = (Z_\lambda^\tau)^{\underline{k}^\#} + V$. It is fairly easy to see that $\underline{k}^\#\cdot V = V$. So, $0 = \langle \underline{k}^\#\cdot(Z_\lambda^\tau)^{\underline{k}^\#}, V\rangle = \langle (Z_\lambda^\tau)^{\underline{k}^\#}, \underline{k}^{\#s}\cdot V\rangle$ = {since $(\underline{k}^\#)^t = (\overline{\underline{k}^\#})$ = (also) the complex conjugate of $\underline{k}^\#$} $\langle (Z_\lambda^\tau)^{\underline{k}^\#}, k^\#\cdot V\rangle = \langle (Z_\lambda^\tau)^{\underline{k}^\#}, V\rangle$. Thus $(Z_\lambda^\tau)^{\underline{k}^\#}$ is orthogonal to V, and so by non-singularity, the bilinear form $\neq 0$ on $(Z_\lambda^\tau)^{\underline{k}^\#}$.

Thus we can fix $z_\lambda^\tau \in (Z_\lambda^\tau)^{\underline{k}^\#}$ such that $\langle z_\lambda^\tau, z_\lambda^\tau\rangle = 1$. Recall $u\cdot z_\lambda^\tau = p_u^\tau(\lambda) z_\lambda^\tau$ for u in $U^{\underline{k}}$. So, since $\langle u\cdot z_\lambda^\tau, z_\lambda^\tau\rangle = \langle z_\lambda^\tau, u^s\cdot z_\lambda^\tau\rangle$, $\langle p_u^\tau(\lambda) z_\lambda^\tau, z_\lambda^\tau\rangle = \langle z_\lambda^\tau, p^\tau_{u^s}(\lambda) z_\lambda^\tau\rangle$, and thus $\overline{p_u^\tau(\lambda)} = p^\tau_{u^s}(\lambda)$ for u in $U^{\underline{k}}$. Now we have

Lemma 6.2. For u in $U^{\underline{k}}$, $\overline{p^\tau_{u^s}(\lambda)} = p_u^\tau(\overline{\lambda'})$. Proof: $u - p_u \in \underline{k}^\# U + U\underline{n}$. So, $\bar{u} - \overline{p_u^\tau} \in \overline{\underline{k}^\#} U + U\underline{n}$. So, applying $p^{-\tau}$, $p^{-\tau}_{\bar{u}} = \overline{p_u^\tau}$. On the other hand, from Lemma 3.4, we get, for all v in $U^{\underline{k}}$, $p^\tau_{v^t}(\lambda) = p_v^{-\tau}(\lambda')$. Setting $v = \bar{u}$, $p^\tau_{u^s}(\lambda) = p^{-\tau}_{\bar{u}}(\lambda')$=(by the above) $\overline{p_u^\tau}(\lambda')$. Taking conjugates, we get $\overline{p^\tau_{u^s}(\lambda)} = \overline{\overline{p_u^\tau}(\lambda')} = p_u^\tau(\overline{\lambda'})$.

Q.E.D.

Now, by Theorem 3.1, $p^\tau\colon U^{\underline{k}} \to U(\underline{a})^{W_\tau}$ is onto. Thus $p_u^\tau(\mu_1-\rho) = p_u^\tau(\mu_2-\rho)$ for all u in $U^{\underline{k}}$ if and only if $\mu_1 \sim_W \mu_2$. So, $p_u^\tau(\overline{\lambda'}) = p_u^\tau(\overline{\mu'}+\rho)$ (since $\bar{\rho}' = \rho$) $= p_u^\tau(-\bar{\mu}+\rho)$ (since $\kappa\mu' = -\mu$) $= p_u^\tau(\lambda)$ if and only if $\mu \sim_W -\bar{\mu}$. So, from Lemma 6.2 we have:

Corollary 6.3. $\overline{p^\tau_{u^s}(\lambda)} = p_u^\tau(\lambda)$ for all u in $U^{\underline{k}}$ if $\mu \sim_W -\bar{\mu}$.

But if $\langle,\rangle$ exists, we have shown $\overline{p^\tau_{u^s}(\lambda)} = p_u^\tau(\lambda)$. Thus $\mu \sim_W -\bar{\mu}$.

Now, if u,v are in U, $\langle u\cdot z_\lambda^\tau, v\cdot z_\lambda^\tau\rangle = \langle (v^s u)_0 z_\lambda^\tau, z_\lambda^\tau\rangle = p^\tau_{(v^s u)_0}(\lambda)$. Thus, $\langle,\rangle$ is unique up to a scalar if it exists.

Now, conversely, suppose $\mu \sim_W -\bar{\mu}$. We want to show $\langle u\cdot z_\lambda^\tau, v\cdot z_\lambda^\tau\rangle = p^\tau_{(v^s u)_0}(\lambda)$ is a U-invariant Hermitian form. First note that if $u\cdot z_\lambda^\tau = u'\cdot z_\lambda^\tau$, $v\cdot z_\lambda^\tau = v'\cdot z_\lambda^\tau$, then $p^\tau_{(v^s u)_0}(\lambda) = p^\tau_{(v'^s u')_0}(\lambda)$. Also, clearly, this form is linear in the first variable. Conjugate linearity follows from $\langle u\cdot z_\lambda^\tau, v\cdot z_\lambda^\tau\rangle = \overline{p^\tau_{(v^s u)_0}(\lambda)}$ = {since $(v^s u)^s = u^s v$ and $U^{\underline{k}}$ and $[\underline{k},U]$ are closed under $u \to u^s$} $\overline{p^\tau_{(u^s v)_0{}^s}(\lambda)}$ = (by Corollary 6.3, since $\mu \sim_W -\bar{\mu}$) $p^\tau_{(u^s v)_0}(\lambda) = \langle v\cdot z_\lambda^\tau, u\cdot z_\lambda^\tau\rangle$.

Finally, we obtain U-invariance. Let $q \in U$. Then $\langle q\cdot(u\cdot z_\lambda^\tau), v\cdot z_\lambda^\tau\rangle = \langle qu\cdot z_\lambda^\tau, v\cdot z_\lambda^\tau\rangle = p^\tau_{(v^s qu)_0}(\lambda) = p^\tau_{((q^s v)^s u)_0}(\lambda) = \langle u\cdot z_\lambda^\tau, q^s v\cdot z_\lambda^\tau\rangle = \langle u\cdot z_\lambda^\tau, q^s\cdot(v\cdot z_\lambda^\tau)\rangle$.

Q.E.D.

Now, by Nelson's theorem, Z_λ^τ can be extended to a unitary representation of G_0 if and only if there exists a positive definite U-invariant Hermitian form on Z_λ^τ (see section 14 for details). Let $\underline{a}^* = \{\lambda \in \underline{a}' \mid \lambda = \mu-\rho,\ \mu_{\sim W}-\bar{\mu}\}$. Let $a_+^* = \{\lambda$ in $\underline{a}^*$ such that Z_λ^τ admits a positive definite U-invariant Hermitian form$\}$. Recall the matrix R_τ^γ from section 5. We will prove:

Theorem 6.4. We have $\lambda \in \underline{a}_+^*$ if and only if $\lambda \in \underline{a}^*$ and $R_\tau^\gamma(\lambda)$ is positive semidefinite. Proof: Let $\gamma \in \Gamma$. Then $\gamma \circ \mathrm{Ad} \in \Gamma_0$. Since $K_R = \mathrm{Ad}(K_0)$ is compact, there exists a positive definite K_R-invariant Hermitian form on V_γ, and thus also a K_0-invariant form B on V_γ (writing γ for the representation $\gamma \circ \mathrm{Ad}$ of K_0) which is still positive definite. Schur's lemma and the irreducibility of V_γ imply its uniqueness up to positive real scalar.

Now for $v \in V_\gamma$, we define v^s in $V_\gamma{}'$ by $\langle v^s,w\rangle = B(w,v)$ for all v in V_γ. Since $(V_\gamma{}')' = V_\gamma$, $v \to v^s$ also defines a map from $V_\gamma{}'$ to V_γ.

Now it is easy to see that H^* is closed under conjugation and transpose. Recall for u in U, $u^s = \bar{u}^t$. Define a map $\sigma \to \sigma^s$: $E_\gamma \to E_\gamma{}'$, by $\sigma^s(v') = (\sigma(v'^s))^s$ for all v' in $V_\gamma{}'$. Let $\{\sigma_1,\ldots,\sigma_{l(\gamma)}\}$ be a basis of E_γ; then $\{\sigma_1{}^s,\ldots,\sigma_{l(\gamma)}{}^s\}$ is a basis for $E_\gamma{}'$. Similarly, if $\{v_1,\ldots,v_r\}$ is a basis for V_γ, $\{v_1{}^s,\ldots,v_r{}^s\}$ is a basis of $V_\gamma{}'$. We now show that, with respect to these bases, $R_\tau^\gamma(\lambda)$ is Hermitian. For, $(R_\tau^\gamma)_{ij}(\lambda) = p^\tau_{z(i,j)}(\lambda) = p^\tau_{z(\sigma_i{}^s,\sigma_j)}(\lambda) =$

$$p^\tau_{\sum_{m=1}^{\ell(\gamma)} \sigma_i{}^s(v_m{}^s)\sigma_j(v_m)}(\lambda) = p^\tau_{\sum_{m=1}^{\ell(\gamma)} (\sigma_j(v_m)^s\sigma_i(v_m))^s}(\lambda) = \text{(by Corollary 6.3)}$$

$$\overline{p^\tau_{\sum_{m=1}^{\ell(\lambda)} \sigma_j(v_m)^s\sigma_i(v_m)}(\lambda)} = \overline{p^\tau_{z(j,i)}(\lambda)}.$$ Thus $R_\tau^\gamma(\lambda)$ is Hermitian, and thus diagonalizable with real eigenvalues. If L is such that $L^{-1}(R_\tau^\gamma)L$ is a real diagonal matrix, and $\sigma_i{}' = \sum_{k=1}^{\ell(\gamma)} L_{ki}\sigma_k$, then it is routine to show that with respect to the bases $\{\sigma_i{}'\}$ and $\{\sigma_i'^s\}$, $R_\tau^\gamma(\lambda)$ is a real diagonal matrix. So, without loss of generality, we assume $\{\sigma_i\}$ is the basis $\{\sigma_i{}'\}$ above, that is, $p_{z(i,j)} = 0$ if $i \neq j$.

Now, the center of G_0 acts trivially on U, so $U^{\underline{k}} = U^K = U^{K_0}$. Recall the map $u \to u_0$: $U \to U^{\underline{k}}$. If $\sigma \in E_\gamma$, $\nu \in E_\gamma{}'$, $(\nu(V_\gamma{}')\sigma(V_\gamma))^{K_0}$ is 1-dimensional by Schur's lemma. Thus, if $v \in V_\gamma$, $v' \in V_\gamma{}'$, $(\nu(v')\sigma(v))_0$ is a scalar multiple of $z(\nu,\sigma)$, since both are K_0-fixed. So, $p^\tau_{\nu(v')\sigma(v)_0}$ is a scalar multiple of $p^\tau_{z(\nu,\sigma)}$. Now if u, $v \in V_\gamma$, $\langle\sigma_i(u)z_\lambda^\tau,\ \sigma_j(v)z_\lambda^\tau\rangle = \langle\sigma_j(v)^s\sigma_i(u)z_\lambda^\tau, z_\lambda^\tau\rangle$. But $\sigma_j(v)^s\sigma_i(u)z_\lambda^\tau - (\sigma_j(v)^s\sigma_i(u))_0 z_\lambda^\tau \in \sum_{\gamma\neq\text{iden.}} (\tau \otimes Z_\lambda^\tau)_\gamma$ by Proposition 4.9. But $(Z_\lambda^\tau)^{\underline{k}^\#} = (\tau \otimes Z_\lambda^\tau)^{K_0}$ is orthogonal to $\sum_{\gamma\neq\text{iden.}} (\tau \otimes Z_\lambda^\tau)_\gamma$ by the first part of the proof of Theorem 6.1. Thus, $\langle\sigma_i(u)z_\lambda^\tau,\ \sigma_j(v)z_\lambda^\tau\rangle = \langle(\sigma_j(v)^s\sigma_i(u))_0 z_\lambda^\tau, z_\lambda^\tau\rangle = p^\tau_{(\sigma_j(v)^s\sigma_i(u))_0}(\lambda) =$ (on the other hand) $c\,p^\tau_{z(k,j)}(\lambda)$ for some scalar c in R. And since R_τ^γ is diagonal by our choice of basis, this expression = 0 if $i \neq j$. So, $\langle\sigma_i(V_\gamma)z_\lambda^\tau,\ \sigma_j(V_\gamma)z_\lambda^\tau\rangle = 0$ if $i \neq j$. Thus $\sigma_i(V_\gamma)Z_\lambda^\tau$ is orthogonal to

$\sigma_j(V_\gamma)Z_\lambda^\tau$ if $i \neq j$.

Now assume $V_{\gamma,i} = \sigma_i(V_\gamma)\cdot z_\lambda^\tau \neq 0$. Then the restriction of $\langle\,,\rangle$ to $V_{\gamma,i}$ is a non-singular K_0-invariant Hermitian form on $V_{\gamma,i}$, and thus a real scalar multiple of the unique K_0-invariant form B defined on V_γ and thus on $V_{\gamma,i}$. So, if $u \in V_\gamma$, $\langle\sigma_i(u)z_\lambda^\tau, \sigma_i(u)z_\lambda^\tau\rangle = cB\langle\sigma_i(u)z_\lambda^\tau, \sigma_i(u)Z_\lambda^\tau\rangle$ where $c \neq 0$, and is independent of u. So, either $\langle\sigma_i(u)z_\lambda^\tau, \sigma_i(u)z_\lambda^\tau\rangle > 0$ or < 0 for all u in V_γ.

Let $\{u_1,\ldots,u_r\}$ be a basis for V_γ. Then $p^\tau_{z(i,i)} = \sum_{k=1}^{r} p^\tau_{(\sigma_i(u_k)s\sigma_i(u_k))_0}$ $(\lambda) = c'\sum_k\langle\sigma_i(u_k)z_\lambda^\tau, \sigma_i(u_k)z_\lambda^\tau\rangle$, $c' > 0$. Thus $p^\tau_{z(i,i)}$ and $\langle\sigma_i(u)z_\lambda^\tau, \sigma_i(u)z_\lambda^\tau\rangle$ are either both > 0 or both < 0. Now, assume $p^\tau_{z(i,i)}(\lambda) \neq 0$ for $i = 1,\ldots,m$, $p^\tau_{z(i,i)}(\lambda) = 0$ for $i = m+1,\ldots,l(\gamma)$. Then by Theorem 5.3, $m = m_\lambda(\gamma)$, and from Lemma 5.4, one sees $\sigma_i(V_\gamma)z_\lambda^\tau = 0$ for $i \geq m+1$, and $\neq 0$ for $i \leq m$. So, assume for $k = 1,\ldots,k$, $p^\tau_{z(i,i)}(\lambda) > 0$, and for $i = k+1,\ldots,m$, $p^\tau_{z(i,i)}(\lambda) < 0$. Now, $(\tau \otimes Z_\lambda^\tau)_\gamma = V_{\gamma,1} + \ldots + V_{\gamma,m_\lambda(\gamma)}$. We see that the restriction of $\langle\,,\rangle$ to $V_{\gamma,i}$ is positive definite for $i = 1,\ldots,k$ and negative definite for $i = k+1,\ldots,m_\lambda$ (γ).

Thus, $\langle\,,\rangle$ is positive definite on $(\tau \otimes Z_\lambda^\tau)$ if and only if $k = m_\lambda(\gamma)$. On the other hand, since R_τ^γ is diagonal, $R_\tau^\gamma(\lambda)$ is positive semidefinite if and only if $p^\tau_{z(i,)} \geq 0$ for all i, that is, since $p^\tau_{z(i,i)} < 0$ for $i = k+1,\ldots,m_\lambda$ (γ), if and only if $k = m_\lambda(\gamma)$. So, $\langle\,,\rangle$ is positive definite on $(\tau \otimes Z_\lambda^\tau)_\gamma$ if and only if $R_\tau^\gamma(\lambda)$ is positive semidefinite, and thus $\langle\,,/$ is positive definite on Z_λ^τ if and only if $R_\tau^\gamma(\lambda)$ is positive semidefinite for all γ in Γ_0.

Q.E.D.

<u>Section 7</u>. An Expression for R_τ^γ.

In this section we prove

<u>Theoren 7.1</u>. For γ in Γ_0, $\lambda \in \underline{a}'$, $R_\tau^\gamma(\lambda) = P_\tau^\gamma(-\bar{\lambda}-2\rho)^* P_\tau^\gamma(\lambda)$, where * denotes conjugate transpose. Proof: There exists (see 9b.) a real bilinear map: $X_\lambda^\tau \times X_{-\lambda-2\rho}^{-\tau} \to C$: $(f,g) = \int_{K_0/M_0} f(a)g(a)da$, which is well defined since $f(km)g(km) = f(k)g(k)$ for k in K_0, m in M_0, f in X_λ^τ, and g in $X_{-\lambda-2\rho}^{-\tau}$. We can also write the integral over the compact space $K_0/(\text{center of } G_0) = Ad(K_0) = K_R$. As usual, we give K_R total measure 1.

If ε is in E_γ, ε' in $E_\gamma{}'$, v in V_γ, v' in $V_\gamma{}'$, then $\left(\varepsilon'(v')^t\varepsilon(v)1_\lambda^\tau, 1_{-\lambda-2\rho}^{-\tau}\right) = \left((\varepsilon'(v')^t\varepsilon(v)\right)_0 1_\lambda^\tau, 1_{-\lambda-2\rho}^{-\tau})$. This follows since $\underline{k}^\#$ annihilates 1_λ^τ and $(\underline{k}^\#)^t = \overline{\underline{k}^\#}$ annihilates $1_{-\lambda-2\rho}^{-\tau}$. So the above expression $= p^\tau_{\{\varepsilon'(v')^t\varepsilon(v)\}_0}$ (λ) since for u in $U^{\underline{k}}$, $u\cdot 1_\lambda^\tau = p_u^\tau(\lambda)1_\lambda^\tau$.

Now let π: $U \to S^*$ be the projection under $U = \underline{k}^\# U + S^*$ (see Corollary 2.6). Then π is a K_0-linear map and $\pi(U^{\underline{k}}) \subset J^*$. Let B_1, B_2: $V_\gamma{}' \otimes V_\gamma \to S^*$ be defined by $B_1(v' \otimes v) = \langle v',v\rangle\left(z(\varepsilon'^t,\varepsilon)\right)$ and $B_2(v' \otimes v) = \left((\varepsilon'(v')^t\varepsilon(v))_0\right)$. They are both K_0-linear, and their images are both K_0-fixed. Since $(V_\gamma' \otimes V_\gamma)^{K_0}$ is one dimensional, we get $\pi\left((\varepsilon'(v')^t\ (v))_0\right) = c\langle v,v'\rangle\pi\left(z(\varepsilon'^t,\varepsilon)\right)$, c in C.

Letting $\{v_1,\ldots,v_r\}$ be a basis of V_γ, $\{v_1',\ldots,v_r'\}$ a dual basis for V_γ', recall $z(\varepsilon'^t,\varepsilon) = \sum_{i=1}^{r} \varepsilon'(v_i')^t\varepsilon(v_i) = \sum_{i=1}^{r} \left(\varepsilon'(v_i')^t\varepsilon(v_i)\right)_0$ since $z(\varepsilon'^t,\varepsilon)$ is in $\underline{U}^{\underline{k}}$. Applying π to both sides, we get $c = 1/r = 1/(\dim V_\gamma)$. Applying p^τ, and noting that $p^\tau\circ\pi = p$, we get $\frac{\langle v',v\rangle}{\dim V_\gamma} p^\tau_{z(\varepsilon'^t,\varepsilon)}(\lambda) = p^\tau_{\left(\varepsilon'(v')^t\varepsilon(v)\right)_o}(\lambda) = \left(\varepsilon(v)\cdot 1^\tau_\lambda,\ \varepsilon'(v')\cdot 1^{-\tau}_{-\lambda-2\rho}\right)$. Writing $f = \varepsilon(v)\cdot 1^\tau_\lambda$, $g = \varepsilon'(v')\cdot 1^{-\tau}_{-\lambda-2\rho}$, and da = Haar measure, we get $\left(\varepsilon(v)\cdot 1^\tau_\lambda,\ \varepsilon'(v')\cdot 1^{-\tau}_{-\lambda-2\rho}\right) = \int_{K_R} f(a)g(a)da$.

Define δ in $(X^\tau_\lambda)'$ by $\langle\delta,f\rangle = f(e)$. Then $f(a) = (a^{-1}\cdot f)(e) = \langle\delta,a^{-1}\cdot f\rangle = \langle\delta,a^{-1}\cdot\left(\varepsilon(v)\cdot 1^\tau_\lambda\right)\rangle = \langle\delta,\varepsilon(a^{-1}\cdot v)\cdot 1^\tau_\lambda\rangle = \langle(a^{-1}\cdot v)^t\cdot\delta,1^\tau_\lambda\rangle = p^{\varepsilon(a^{-1}\cdot v)}_\tau(\lambda)$. Recall $p^u_\tau(\lambda) = p^{-\tau}_u(\lambda')$. The last equality holds because $\langle(\underline{k}^{\#}U + U\underline{n})\cdot\gamma,1^\tau_\lambda\rangle = 0$. Similarly, $g(a) = p^{\varepsilon'(a^{-1}\cdot v)}_{-\tau}(-\lambda-2\rho)$.

Thus, $\left(\varepsilon(v)\cdot 1^\tau_\lambda,\ \varepsilon'(v')\cdot 1^{-\tau}_{-\lambda-2\rho}\right) = \int_{K_R} f(a)g(a)da = \int_{K_R} p^{\varepsilon(a^{-1}\cdot v)}_\tau(\lambda)\cdot p^{\varepsilon'(a^{-1}\cdot v)}_{-\tau}(-\lambda-2\rho)\,da$. Now define T_λ: $E_\gamma \to V_\gamma'$ and T_λ': $E_\gamma' \to V_\gamma$: $\langle T_\lambda(\varepsilon),v\rangle = p^{\varepsilon(v)}_\tau$ for all v in V_γ and $\langle v',T_\lambda'(\varepsilon')\rangle = p^{\varepsilon'(v')}_{-\tau}(\lambda)$ for all v' in V_γ'. So, $\left(\varepsilon(v)\cdot 1^\tau_\lambda,\ \varepsilon'(v')\cdot 1^{-\tau}_{-\lambda-2\rho}\right) = \int_{K_R} p^{\varepsilon(a^{-1}v)}_\tau(\lambda)p^{\varepsilon'(a^{-1}v')}_{-\tau}(-\lambda-2\rho)da = \int_{K_R}\langle T_\lambda(\varepsilon),a^{-1}v\rangle\langle a^{-1}v',T'_{-\lambda-2\rho}(\varepsilon')\rangle\,da$ = (by chapter 4, 9c.) $(\dim V_\gamma)^{-1}\cdot\langle T_\gamma(\varepsilon),T_{-\lambda-2\rho}(\varepsilon')\rangle\langle v',v\rangle$. But we calculated earlier that the original expression in these equations = $(\dim V_\gamma)^{-1}\langle v',v\rangle\, p^\tau_{z(\varepsilon'^t,\varepsilon)}(\lambda)$.

Now, if $a\in M_o$, v' in V_γ', $\langle v',a\,T_\lambda'(\varepsilon')\rangle = \langle a^{-1}v',T_\lambda'(\varepsilon')\rangle = p^{\varepsilon'(a^{-1}v')}_{-\tau}(\lambda) = p^{a^{-1}\cdot\varepsilon(v')}_{-\tau}(\lambda)$ = (since $p_{-\tau}$ is M_0-linear) $p^{\varepsilon'(v')}_{-\tau}(\lambda) = \langle v',T_\lambda'(\varepsilon')\rangle$. Thus, $T_\lambda'(\varepsilon')$ is M_0-fixed. Similarly, $T_\lambda(\varepsilon)$ is M_0-fixed.

Now choose a K_0-invariant positive definite Hermitian form on V_γ, and let $\{v_1,\ldots,V_{1(\gamma)}\}$ be an orthonormal basis of $(V_\gamma)^{M_0}$. Then $\{v_1^s,\ldots,v_{1(\gamma)}^s\}$ is a basis of $(V_\gamma')^{M_0}$, where $\langle v^s,v_0\rangle = B(v_0,v)$, B the Hermitian form, and v,v_o in $(V_\gamma)^{M_0}$.

Now, $T_\lambda(\varepsilon) = \sum_{i=1}^{\ell(\gamma)}\langle T_\lambda(\varepsilon),v_i\rangle v_i^s$, and $T'_{-\lambda-2\rho}(\varepsilon') = \sum_{i=1}^{\ell(\gamma)}\langle v_i^s,T'_{-\lambda-2\rho}(\varepsilon')\rangle\cdot v_i$. Recall $p^\tau_{z\left((\varepsilon')t,\varepsilon\right)}(\lambda) = \langle T_\lambda(\varepsilon),T'_{-\lambda-2\rho}(\varepsilon')\rangle$ = (by the above) $\sum_{i=1}^{\ell(\gamma)}\langle T_\lambda(\varepsilon),v_i\rangle\langle v_i^s,T'_{-\lambda-2\rho}(\varepsilon')\rangle$.

Choose a basis $\{\varepsilon_1,\ldots,\varepsilon_{1(\gamma)}\}$ of E_γ, and a basis $\{\varepsilon_1^s,\ldots,\varepsilon_{1(\gamma)}^s\}$ of E_γ' such that $\varepsilon_i^s(v') = \left(\varepsilon_i(v'^s)\right)^s$. Substituting $\varepsilon = \varepsilon_j$, $\varepsilon' = (\varepsilon_k^s)^t$ in the last expression, $p^\tau_{z(\varepsilon_k^s,\varepsilon_j)}(\lambda) = \left(R^\gamma_\tau(\lambda)\right)_{kj} = \sum_{i=1}^{\ell(\gamma)}\langle T_\lambda(\varepsilon_j),v_i\rangle\langle v_i^s,T'_{-\lambda-2\rho}(\varepsilon_k^s)^t\rangle = \Sigma p_\tau^{\varepsilon_j(v_i)}(\lambda)p_{-\tau}^{(\varepsilon_k^s)^t(v^s)}(-\lambda-2\rho) = p_\tau^{\varepsilon_j(v_i)}(\lambda)p_{-\tau}^{\overline{k(v_i)}}(-\lambda-s\rho)$. This follows since $(\varepsilon_k^s)^t(v_i^s) = (\varepsilon_k^t)^s(v_i^s) = \varepsilon_k^t(v_i)^s = \overline{\varepsilon_k(v_i)}$. Now, $u - p_u\varepsilon(\underline{k}^{\#}U+U\underline{n})$. Thus, $\overline{u} - \overline{p_u}\ \varepsilon\ (\underline{k}^{\#} + U\underline{n})$. So, we have $\overline{p^{-\tau}_u} = \overline{p^\tau_u}$, and thus $\overline{p^u_\tau} = p^{\overline{u}}_{-\tau}$.

Thus the expression above $= \sum_{i=1}^{\ell(\gamma)} p_\tau^{\varepsilon_j(v_1)}(\lambda)\ \overline{p^{\varepsilon k(v_i)}(-\lambda 2\rho)} = \sum_{i=1}^{\ell(\gamma)} p_\tau^{\varepsilon_j(v_i)}(\lambda)\ \overline{p_\tau^{\varepsilon_j(v_i)}(-\overline{\lambda}-2\rho)} = P^\gamma_\tau(-\overline{\lambda}-2\rho)^*P^\gamma_\lambda(\lambda)_{kj} = (R^\gamma_\tau)_{kj}$.

Q.E.D.

Chapter III

Section 8. Reduction to Rank One.

Let $\gamma \in \Gamma$. Let $\{\varepsilon_1,\ldots,\varepsilon_{l(\gamma)}\}$ be a basis of E_γ, $\{v_1,\ldots,v_{l(\gamma)}\}$ a basis for $V_\gamma{}^M$. Let P_γ^τ be the $U(\underline{a})$-valued matrix $(P_\gamma^\tau)_{ij} = p^\tau_{\varepsilon_i(v_j)}$. We show in section 13 that $p_\gamma^\tau = \det P_\gamma^\tau$ is a non-zero scalar multiple of $\det P_\tau^\gamma$. In this section we evaluate p_γ^τ as a product of similar expressions corresponding to rank one subgroups.

Recall from section 2 that for each $\emptyset$ in $\Delta_+{}^1$, we obtain groups $G_\emptyset$ and $K_\emptyset$, and elements $\{\delta_1{}^\emptyset,\ldots,\delta^\emptyset_{l(\gamma)}\}$ in $\hat{K}_\emptyset$. Recall the polynomials $p_\emptyset^\tau = p^\tau_{\delta^\phi}\cdots p^\tau_{\delta^\phi_{\ell(\gamma)}}$, and $p^\tau_{(\emptyset)} = T_{\rho_\phi-\rho}(p_\emptyset)$. In this section we will prove:

Theorem 8.1. Up to a scalar multiple, $p_\gamma^\tau = \pi_{\phi\varepsilon\Delta_+{}^1} P(\emptyset)$.

Now let α be a simple root. Let $\underline{n}^\alpha = \sum_{\emptyset\varepsilon\Delta_+-\{\alpha\}} (\underline{g}^\emptyset + \underline{g}^{2\emptyset})$. Clearly $(\text{ad }\underline{g})\cdot(\underline{n}^\alpha) \subset \underline{n}^\alpha$. Let $\underline{k}_\alpha$ and $\underline{n}_\alpha$ play the roles of $\underline{k}$ and $\underline{n}$ for the subalgebra $\underline{g}_\alpha$, J_α^* and H_α^* the roles of J^* and H^*. We first prove:

Proposition 8.2. The map $J_\alpha^* \otimes H_\alpha^* \otimes U(\underline{n}^\alpha) \to U$ given by $u \otimes v \otimes w \to uvw$ is an injection. Identifying this space with its image in U we get $U = \underline{k}^\# U + J_\alpha^* \otimes H_\alpha^* \otimes U(\underline{n}^\alpha) = \underline{k}^\# U + J_\alpha^* H_\alpha^* + J_\alpha^* H_\alpha^* U(\underline{n}^\alpha)\underline{n}^\alpha$. Proof: The proof of this proposition is very similar to that of Proposition 2.4.1 in ICSR. It is clear that $\underline{K}^\# U$ is a right ideal in U. Let $f\colon U \to U/\underline{k}^\#$ be the natural map. It is sufficient to prove that the map $J_\alpha^* \otimes H_\alpha^* \otimes U(\underline{n}^\alpha) \to U/\underline{k}^\# U$ given by $u \otimes v \otimes w \to f(uvw)$ is a bijection.

Let $\underline{s} = \underline{a} + \underline{n}$, $\underline{s}_\alpha = \underline{a} + \underline{n}_\alpha$. So, $\underline{s} = \underline{s}_\alpha + \underline{n}^\alpha$. Note that $U(\underline{s}) = U(\underline{s}_\alpha) \otimes U(\underline{n}^\alpha)$. Now, one can see that $U = \underline{k}^\# U + U(\underline{s})$. Thus f is a vector space isomorphism from $U(\underline{s})$ to $U/\underline{k}^\# U$. Now, $U/\underline{k}^\# U$ is a right U-module, and if $u,v \in U$, $f(uv) = f(u)\cdot v$. Thus $f_1\colon U(\underline{s}_\alpha) \otimes U(\underline{n}^\alpha) \to U/\underline{k}^\# U$, defined by $f_1(u \otimes v) = f(uv)$, is an isomorphism.

By Corollary 2.6, $U(\underline{g}_\alpha) = \underline{k}_\alpha{}^\# U(\underline{g}_\alpha) + U(\underline{s}_\alpha) = \underline{k}_\alpha{}^\# U(\underline{g}_\alpha) + J_\alpha^* H_\alpha^*$. Thus there is a bijection $P\colon J_\alpha^* H_\alpha^* \to U(\underline{s}_\alpha)$ such that $u - P(u)$ is in $\underline{k}_\alpha{}^\# U(\underline{g}_\alpha)$ if u is in $J_\alpha^* H_\alpha^*$. So, we have an isomorphism $f_2\colon J_\alpha^* H_\alpha^* \otimes U(\underline{n}^\alpha) \to U/\underline{k}^\# U$ defined by $f_2(u \otimes v) = f_1\big(P(u) \otimes v\big)$. Now, $f_1\big(P(u) \otimes v\big) = f\big(P(u)\big)\cdot v = f(u)\cdot v$, and so $f_2(u \otimes v) = f(uv)$. But the product map: $J_\alpha^* \otimes H_\alpha^* \to J_\alpha^* H_\alpha^*$ is an isomorphism. Thus the map $J_\alpha^* \otimes H_\alpha^* \otimes U(\underline{n}^\alpha) \to U/\underline{k}^\# U$ is a bijection.

Q.E.D.

Remark 8.3. Let $u \to \pi(u)$ be the projection to $J_\alpha^* H_\alpha^*$ in the decomposition given by the previous proposition. Then $p_u^\tau = p^\tau_{\pi(u)}$. Proof: Note that π is a K_α-linear map, since K_α leaves each summand stable. Also, $U(\underline{a}) \subset J_\alpha^* H_\alpha^*$ and

$\underline{n} \supset \underline{n}^{\alpha}$. The remark follows easily.

Now we have the tools to prove:

Proposition 8.4. We have $p^{\tau}_{\varepsilon_i(v_j)} = p^{\tau}_{\delta^{\alpha}_j}\cdot r_{ij}$, where r_{ij} is invariant under $(W_\alpha)_T$, W_α the Weyl group of $(\underline{g}_\alpha, a)$. Proof: If $u \in U(\underline{g}_\alpha)$, then we can consider p^{τ}_u with respect to either of the 2 decompositions $U = \underline{k}^{\#}U + U(\underline{a}) + U\underline{n}$, and $U(\underline{g}_\alpha) = \underline{k}_\alpha{}^{\#}U(\underline{g}_\alpha) + U(\underline{a}) + U(\underline{g}_\alpha)\underline{n}_\alpha$. But $\underline{k}_\alpha{}^{\#} \subset \underline{k}^{\#}$ and $\underline{n}_\alpha \subset \underline{n}$, so these 2 decompositions are the same.

Now define R_{ij} as the K_α- module generated by $\varepsilon_i(v_j)$. As an abstract K_α-module, $R_{ij} = V_{\delta^{\alpha}_j}$. Thus, recalling π from Remark 8.3, $\pi(R_{ij}) \subset (J^*_\alpha H^*_\alpha)_{\delta^{\alpha}_j}$, that is, the K_α-submodule of $J^*_\alpha H^*_\alpha$ behaving as a multiple of $V_{\delta^{\alpha}_j}$. But J^*_α is K_α-fixed, so $\pi(R_{ij}) \subset J^*_\alpha\cdot(H^*_\alpha)_{\delta^{\alpha}_j}$, and we recall that δ^{α}_j appears uniquely in H^*_α (Theorem 2.16). Thus $\pi(\varepsilon_i(v_j)) = q_{ij}v_j{}'$, where $q_{ij} \in J^*_\alpha$, and $v_j{}'$ is the unique element (up to scalar) in $(H^*_\alpha)_{\delta^{\alpha}_j}$ which is M-fixed. So, $p^{\tau}_{\varepsilon_i(v_j)} = p^{\tau}_{\pi(\varepsilon_i(v_j))} = p^{\tau}_{q_{ij}v_j{}'} =$ (by Lemma 4.2) $p^{\tau}_{q_{ij}}p^{\tau}_{v_j{}'} = p^{\tau}_{q_{ij}}p^{\tau\alpha}_{\delta^{\alpha}_j}$. Now we have only to show that $p^{\tau}_{q_{ij}}$ is invariant under $(W_\alpha)_T$. But this follows from Theorem 3.1, applied to $\underline{g}_\alpha$.

Q.E.D.

Let κ_α be the non-trivial element of W_α, $\tilde{\kappa}_\alpha = T_{-\rho}\kappa_\alpha T_\rho$. Then we easily see that $\tilde{\kappa}_\alpha = T_{-\rho_\alpha}\kappa_\alpha T_{\rho_\alpha}$, and thus lies in $(W_\alpha)_T$. Thus r_{ij} is $\tilde{\kappa}_\alpha$-invariant.

Now, we have proved that $p^{\tau}_{\delta^{\alpha}_j}$ divides every element in the j-th column of P^{τ}_γ, and thus it divides p^{τ}_γ. Thus p_α divides p^{τ}_γ. Note that $p_{(\alpha)} = p_\alpha$.

Proposition 8.5. For $\emptyset$ in $\Delta_+{}^1$, $p_{(\emptyset)}$ divides p_γ. Proof: The proof is almost identical to that of Proposition 2.4.5 of ICSR. If $\emptyset \in \Delta_+{}^1$, $\emptyset = \Sigma m_i\alpha_i$, α_i simple roots. Let $O(\emptyset) = \Sigma m_i$. If $O(\emptyset) = 1$, $\emptyset$ is simple, and we have the result already.

Suppose now that we have proved the result for all $\psi \in \Delta_+{}^1$, such that $O(\psi) < O(\emptyset)$. Then there exists a simple root α such that $(\alpha,\emptyset) > 0$ (using the inner product on $\underline{a}'$ induced by the Killing form), and a positive root ψ with $O(\psi) < O(\emptyset)$, and $\emptyset = \kappa_\alpha\cdot\psi$. Clearly $\psi \neq \alpha$.

We know $p^{\tau}_\gamma = r_\alpha p^{\tau}_\alpha$, where r_α is $\tilde{\kappa}_\alpha$-invariant. Also, by induction, $p^{\tau}_{(\psi)}$ divides p^{τ}_γ. But p^{τ}_α and $p^{\tau}_{(\psi)}$ are relatively prime: the former is a polynomial in $w_\alpha \in [\underline{g}_\alpha,\underline{g}_\alpha] \cap \underline{a}$, the latter in $w_\psi \in [\underline{g}_\psi,\underline{g}_\psi] \cap \underline{a}$. So, $p^{\tau}_{(\psi)}$ divides r_α, and thus $\tilde{\kappa}_\alpha\cdot p^{\tau}_{(\psi)}$ divides $\tilde{\kappa}_\alpha\cdot r_\alpha = r_\alpha$. The proposition now follows from:

Lemma 8.6. We have $\tilde{\kappa}_\alpha\cdot p^{\tau}_{(\psi)} = c\cdot p^{\tau}_{(\emptyset)}$, c a scalar. Proof: $\kappa_\alpha\cdot\psi = \emptyset$. So, $\kappa_\alpha\cdot\underline{g}_\psi = \underline{g}_\emptyset$, $\kappa_\alpha\cdot\underline{k}_\psi = \underline{k}_\emptyset$, $\kappa_\alpha\cdot\underline{n}_\psi = \underline{n}_\emptyset$, and if $X \in \underline{k}_\psi$, $\kappa_\alpha\cdot(X - \tau(X)) = \kappa_\alpha\cdot X - \tau(\kappa_\alpha\cdot X)$, because any character on k is constant on orbits of Ad(K). So, $\kappa_\alpha\cdot\underline{k}_\psi{}^{\#} = \underline{k}_\emptyset{}^{\#}$. Also, $\kappa_\alpha K_{\kappa_\alpha}{}^{-1} = K_\emptyset$, and $\kappa_\alpha M\kappa_\alpha{}^{-1} = M$.

So, $\kappa_\alpha\cdot V_\gamma{}^M = V_\gamma{}^M$, and κ_α maps the K_ψ-module generated by $V_\gamma{}^M$ to the $K_\emptyset$-module generated by $V_\gamma{}^M$. Further, if the former has k_ψ-decomposition

$\Sigma W_\delta \psi_j$, then letting $W_\delta \phi_j = \kappa_\alpha \cdot W_\delta \psi_j$, the latter has $K_\emptyset$-decomposition $\Sigma W_\delta \phi_j$. So, $\kappa_\alpha \cdot p_\psi^\tau = c \cdot p_\emptyset$. Now, $\tilde{\kappa}_\alpha P_{(\psi)} = T_{-\rho}\kappa_\alpha T_\rho T_{(\rho_\psi - \rho)} P_\psi = T_{-\rho}\kappa_\alpha \cdot T_{\rho_\psi} P_\psi = T_{-\rho}\kappa_\alpha T_{\rho_\psi}\kappa_\alpha^{-1}\kappa_\alpha P_\psi = T_{-\rho} T_{\kappa_\alpha \cdot \rho_\psi} \kappa_\alpha P_\psi = T_{-\rho} T_{\rho_\emptyset} P_\emptyset = T_{(\rho_\emptyset - \rho)} P_\emptyset = P_{(\emptyset)}$.

Q.E.D.

To prove Theorem 8.1, we need only show:

Lemma 8.7. $\pi_{\phi \varepsilon \Delta_+^1} P_{(\emptyset)}$ and p_γ^τ have the same degree. Proof: We need only show p_γ^τ and p_γ^0 have the same degree, because then $p_{\delta\phi_j}^\tau$ and $p_{\delta\phi_j}^0$ will have the same degree, and thus $p_{(\emptyset)}^\tau$ and $p_{(\emptyset)}^0$ will have the same degree, and so p_γ^τ and $\pi_{\phi \varepsilon \Delta_+^1} P_{(\emptyset)}$ will have the same degree. Now, Theorem 2.3.18 of ICSR states that $p_\gamma^0 \neq 0$.

Let u in U be homogeneous. Thus one term of u is $u_0 = X_1 \cdots X_r H_1 \cdots H_s Y_1 \cdot \cdots Y_t$, where $x_i \; \varepsilon \; \underline{k}$, $H_i \; \varepsilon \; \underline{a}$, and $Y_i \; \varepsilon \; \underline{n}$. Observe that if $t > 0$, $p_{u_0}^\tau = p_{u_0}^0 = 0$. If $t = 0$, $p_{u_0}^\tau = \tau(X_1)\cdots\tau(X_r)(H_1)\cdots(H_s)$, while $p_{u_0}^0 = 0$ unless $r = 0$, in which case $p_{u_0}^0 = H_1 \cdots H_s$. Now, if u is homogeneous of degree n, then it is a sum of terms of the form above, where $r + s + t = n$. So, observe that $p_u^0 = 0$, or is a homogeneous polynomial of degree n. In the latter case, note that $\deg p_u^\tau = n$ also, though p_u^τ need not be homogeneous. Further, in this case, $\deg (p_u^\tau - p_u^0) < n$.

So, since $\varepsilon_i(v_j)$ is homogeneous of some degree $d_i(\gamma)$ (this is shown in 5a.), we know that either $p_{\varepsilon_i(v_j)}^0$ is 0, or else $p_{\varepsilon_i(v_j)}^\tau$ has degree $d_i(\gamma)$. But at least one element in each column of P_γ^0 is non-0, since $p_\gamma^0 \neq 0$, so at least one element of each column of P_γ^τ has degree $d_i(\gamma)$, and thus $\deg p_\gamma^\tau = \Sigma d_i(\gamma) = \deg p_\gamma^0$.

Q.E.D.

Thus we have shown that the 2 polynomials in Theorem 8.1 have the same degree, and one divides the other, so they are equal up to scalar.

Section 9. The Rank One Case.

Let $\underline{g}$ be a rank one reductive Lie algebra. We wish to find p_γ^τ as γ ranges through Γ. In classifying non-compact symmetric spaces, E. Cartan has shown that in the rank one case $\underline{k}$ is semisimple unless $[\underline{g},\underline{g}]_R = su(n,1)$, $n \geq 1$ (see 2a., chapter 9). The character τ is thus trivial except in this case. In light of the previous section, the important example is $\underline{g} = \underline{g}_\emptyset$, $\emptyset$ in Δ_+^1.

Recall Theorem 2.16, that if rank $\underline{g} = 1$, $l(\gamma) = 1$. Thus if $0 \neq \varepsilon$ in E_γ, $0 \neq v \; \varepsilon \; V_\gamma^{M_\theta}$, then $p_\gamma^\tau = p_{\varepsilon(v)}^\tau$ up to scalar.

In this section, we summarize results from ICSR. An element $z \; \varepsilon \; \underline{k}$ is found, conjugate to $-w$, along with a space $\underline{d}$, which is the +1 eigenspace of z in $\underline{p}$. A parametrization $\{\gamma_{ij}\}$ of Γ is found, along with $u_{ij} \; \varepsilon \; S(\underline{d})$, such that $p_{\gamma_{ij}}^0 = p_{\beta(u_{ij})}^0$, where β is the symmetrization map. We will show that, in fact, $p_{\gamma_{ij}}^\tau = p_{\beta(u_{ij})}^\tau$.

Recall that the Killing form on $[\underline{g},\underline{g}]$: $(X,Y) = \text{tr}(\text{ad } X \text{ ad } Y)$ is negative definite on $\underline{k}_R$ and can be extended to all of $\underline{g}$ so that it is positive definite on $\underline{p}_R$. Thus $B(X,Y) = -(X,\theta Y)$ is positive definite on $\underline{g}_R$ (θ is the Cartan involution).

Let $\emptyset$ be the unique element in Δ_+^1, w in $\underline{a} \cap [\underline{g},\underline{g}]$ such that $\emptyset(w) = 1$. Let $e_1 \in \underline{g}^{\emptyset}$ be such that $B(e_1,e_1) = \frac{1}{2}B(w,w)$. Let $e_{-1} = -\theta e_1$. Then, w, e_1, and e_{-1} span a Lie subalgebra $\underline{u}_R$ of $\underline{g}_R$ isomorphic to $sl(2,R)$. Let $z = i(e_1 - e_{-1}) \in i\underline{k}_R$. Then z is conjugate to $-w$ under $\exp \text{ad}\left(\frac{1}{2}i(e_1 + e_{-1})\right)$. Obviously, the eigenvalues of ad z are those of ad$(-w)$: 0, ± 1, and possibly ± 2. One can see that z acts on $\underline{p}$, and the only eigenvalues of ad z on $\underline{p}$ are 0 and ± 1. Let $\underline{d}$ by the $+1$ eigenspace of ad z on $\underline{p}$. Let $s = \dim \underline{g}^{2\emptyset}$. Then $\dim \underline{d} = s + 1$.

Now, we can decompose $\underline{g}$ as a sum of $\underline{u}$-submodules. Each irreducible $\underline{u}$-submodule of $\underline{g}$ has dimension 1, 3, or 5. Let $\underline{h}$ be the orthogonal complement of $(\underline{a} + \underline{u})$ in $\underline{g}$ under the positive definite form B. Then $\underline{h}$ is clearly a $\underline{u}$-module. Let $\underline{h}_i$ be the span of the irreducible $\underline{u}$-submodules of $\underline{h}$ of dimension $(2i + 1)$. So, $\underline{h} = \underline{h}_0 + \underline{h}_1 + \underline{h}_2$. Recalling $\underline{g} = \sum_{j=-2}^{+2} \underline{g}^{j\emptyset}$, let $\underline{h}_i{}^j = \underline{h}_i \cap \underline{g}^{j\emptyset}$. Now, $\underline{g}^0 \cap \underline{h} = \underline{m}$, so call $\underline{h}_i{}^0 = \underline{m}_i$.

Now, note that since the only eigenvalue of z on $\underline{d}$ is $+1$, K^z, the centralizer of z in K, acts on $T = S(\underline{d})$. Let Γ^z be the set of irreducible K^z-representations occurring in T. It is shown in ICSR (Theorem 2.1.28) that every representation of K^z occurs at most once in T. We have the following (a combination of Theorem 2.2.4 and Lemma 2.2.6 of ICSR):

Theorem 9.1. Let $V \subset H$ be an irreducible K-module. Let j be the highest eigenvalue of z on V. Let $V_z = V \cap T_j$ (T_j = homogeneous polynomials of degree j in T). Then $V_z \neq 0$. The correspondence $V \to V_z$ gives a bijection: $\Gamma \to \Gamma^z$.

Now, let M_0 be the analytic subgroup of G with Lie algebra $\underline{m}_0$. ICSR shows that if $\gamma_z \in \Gamma^z$, $(T_{\gamma_z})^{M_0}$ is one dimensional, and that if $u \in (H^*_\gamma)^{M_\theta}$, $v \in (T^*_{\gamma_z})^{M_0}$, then $p_u^0 = p_v^0$ up to scalar.

Now, by Remark 2.2.13 of ICSR, $(T_{\gamma_z})^{M_0} \not\subset (H_\gamma^{M_\theta})^\perp$, that is, the perpendicular complement of $(H_\gamma)^{M_\theta}$ is an M_θ-module. Further, since p^τ is M_θ-linear, and $U(\underline{a})$ is M_θ-fixed, p^τ is 0 on $(H_\gamma^{*M_\theta})^\perp$. Now, if $u \in (T^*_{\gamma_z})^{M_0}$, $u = u_1 + u_2$, $u_1 \in (H_\gamma^*)^{M_\theta}$, $u_2 \in (H^*_\gamma{}^{M_\theta})^\perp$, and $p_u^\tau = p_{u_1}^\tau + p_{u_2}^\tau = p_{u_1}^\tau$. And, in fact, if $u \neq 0$, $u_1 \neq 0$. Thus if v lies in the 1-dimensional subspace $(H_\gamma^*)^{M_\theta}$, and u lies in the 1-dimensional subspace $(T^*_{\gamma_z})^{M_0}$, then $p_u^\tau = p_v^\tau$ up to a scalar multiple. Thus we need only find $p^\tau_{v_{\gamma_z}}$ for v_{γ_z} in $(T^*_{\gamma_z})^{M_0}$. With a few definitions, we will be able to give the structure of T as a K^z-module.

Recall that each element in $S(\underline{p})$ can be considered as a differential operator on $S(\underline{p})$ by $u \to D_u$ for u in $S(\underline{p})$, where if $X,Y \in \underline{p}$, $D_X Y = B(X,Y)$.

Let q be the element in T_2 (unique up to scalar) which is invariant under

$S(\underline{d})$. Then if $\{v_0,\ldots,v_s\}$ are an orthonormal basis for $\underline{d}$, $q = v_0^2 + \ldots + v_s^2$. Let $L_j = u$ in T_j such that $D_q u = 0$, that is, the harmonics (in the classical sense) of degree j. L_j is an irreducible k^z-module, except if $s = 1$, where $L_j = L_j^+ + L_j^-$ as a sum of irreducible modules.

Recall $\underline{g} = \underline{u} + \underline{a} + \underline{h}_0 + \underline{h}_1 + \underline{h}_2$ is the decomposition of $\underline{g}$ into $\underline{u}$-submodules. Let c in G be the element such that $c\cdot(-w) = z$. Then it is shown in section 2.2 of ICSR that $c^{-1}\cdot\underline{h}_2^1 \subset \underline{d}$. Call this space $\underline{b}$. Note dim $\underline{b}$ = dim $\underline{h}_2^1 = s$. Let $O(\underline{b}) \subset SO(\underline{d})$ be the orthogonal matrices fixing the orthogonal complement of $\underline{b}$ in $\underline{d}$. The following is a combination of Theorem 2.1.18 and Remark 2.2.14 of ICSR:

Theorem 9.2. As a direct sum of irreducible K^z-modules, T = $\sum_{i\geq 0, j\geq 0} q^i L_j$ if $s > 1$, $\sum_{i\geq 0,j\geq 0} q^i L_j^+ + \sum_{i\geq 0,j\geq 0} q^i L_j^- + \sum_{i\geq 0} C\cdot q^i$ if $s = 1$, and $\sum_{\substack{i\geq 0\\ j=0 \text{ or } 1}} q^i L_j$ if $s = 0$. Further, $\left(q^i\cdot L_j^{(\pm)}\right)^{M_0} = q^i\cdot\left(L_j^{(\pm)}\right)^{O(\underline{b})}$.

Corollary 9.3. As γ runs through Γ, p_γ^τ runs through $\{p_\beta^\tau\left(q^i\cdot(L_j^{(\pm)})^{O(\underline{b})}\right)\}$.

Now, ICSR determines the elements $q^i\left(L_j^{(\pm)}\right)^{O(\underline{b})}$. In our case, we need only know these elements for $\underline{g}_R = su(N,1)$, $N \geq 1$ (that is, for s = 0 or 1), for otherwise $\underline{k}$ has no center, and the only character on $\underline{k}$ is $\tau = 0$.

Recall c in G such that $c\cdot(-w) = z$. Let $g = c^{-1}\cdot e_1 \in d$. Let $\{q_1,\ldots,q_s\}$ be an orthonormal basis (under $-B$) of $i(\underline{h}_2^1)_R$. Let $h_i = c^{-1}\cdot q_i$. Then, if we normalize the h_i, $\{g,h_1,\ldots,h_s\}$ form a basis for $\underline{d}$, and $q = g_1^2 + h_1^2 +\ldots + h_s^2$.

For the case $\underline{g}_R = sl(2,R)$, see Remark 9.6. For $\underline{g}_R = su(N,1)$, $N \geq 2$, $s = 1$, and dim $\underline{d} = 2$. So, $q = g^2 + h^2$, and it follows directly from the definitions (see Remark 2.1.25 of ICSR) that $(L_j^{\pm})^{O(\underline{b})} = C\cdot(g \pm ih)$.

Now, let us denote multiplication in $S(\underline{p})$ by an $\circ$, and write $x^{(n)}$ for $x\circ\ldots\circ x$, x in $\underline{p}$. Then we wish to calculate $p_\beta^\tau\left((g^{(2)} + h^{(2)})^{(j)} \circ (g\pm ih)^{(k)}\right)$, for j, k in Z^+. Thus it suffices to calculate $p_\beta^\tau\left((g + ih)^{(j)} \circ (g - ih)^{(k)}\right)$ for j, k in Z^+, since $(g + ih) \circ (g - ih) = g^{(2)} + h^{(2)}$. For $\underline{g}_R = su(N,1)$, $N \geq 2$, we now prove:

Lemma 9.4. $[g,h] \in \underline{k}'$, the commutator subalgebra of $\underline{k}$. Proof: Both g and h are in the +1 eigenspace of z in $\underline{p}$. Thus [g,h] is in the +2 eigenspace of z in $[\underline{p},\underline{p}] = \underline{k}$. But $\underline{k} = \underline{k}' + \underline{z}$, where $\underline{z}$ is the center of $\underline{k}$, and thus lies in the 0 eigenspace of ad z in $\underline{k}$. Since $\underline{k}'$ is left stable by ad z, the +2 eignespace of z must lie in $\underline{k}'$. Thus $[g,h] \in \underline{k}'$.

Q.E.D.

Corollary 9.5. $p_\beta^\tau\left((g + ih)^{(j)} \circ (g - ih)^{(k)}\right) = p^\tau(g + ih)^j(g - ih)^k$.

Proof: $[g,[g,h]]$ and $[h,[g,h]]$ both lie in the +3 eigenspace of ad z in $\underline{g}$, which does not exist. Thus, $[g,h]$ commutes with both g and h. Thus we get $\beta\left(g^{(j)}h^{(k)}\right) = g^jh^k$ + terms $\{[g,h]\cdot g^lh^m\}$ with $1 \leq j$, $m \leq k$. Now, $p^\tau_{(\underline{k}'U)} = 0$, and $[g,h] \in \underline{k}'$. Thus $p^\tau_{\beta\left(g^{(j)}h^{(k)}\right)} = p^\tau_{g^jh^k}$. The corollary now follows by linearity.

Q.E.D.

Remark 9.6. If $\underline{g}_R = su(1,1) = sl(2,R)$, the results of Theorems 9.1 and 9.2 do not hold without some modification. Note that in this case, the action of $\underline{k}_R$ on $\underline{p}_R$ is the action of $so(2)$ on R^2.

Now, it is easy to see that the harmonic polynomials in $S(C^2)$ under the action of $SO(2,C)$ are the powers of $(x \pm iy)$, where x and y are orthonormal vectors in C^2. Now, g and $\bar{g}$ (complex conjugate) are orthonormal vectors in $\underline{p}$. Thus as γ runs through Γ, p^τ_γ runs through the sets $\{p^\tau_{g^j}\}$ and $\{p^\tau_{\bar{g}^j}\}$, j in Z^+.

Summarizing,

Corollary 9.7. If $\underline{g}_R = su(1,1)$, then as γ runs through Γ, p^τ_γ runs through $\{p^\tau_{g^j}\}$ and $\{p^\tau_{\bar{g}^j}\}$, j in Z^+. If $\underline{g}_R = su(N,1)$, $N \geq 2$, then as γ runs through Γ, p^τ_γ runs through $\{p^\tau_{(g+ih)^j(g-ih)^k}\}$, j, k in Z^+.

Section 10. The Diagonal Map.

Let $D\colon U \to U \otimes U$ be the unique Lie algebra homomorphism sending X in $\underline{g}$ to $X \otimes 1 + 1 \otimes X$. Results are obtained in ICSR, using D, because the projection map p^0 is well behaved under D. The projection p^τ is not well behaved, however, but there is another projection which is.

Let $\underline{k}'$ be the commutator subalgebra of $\underline{k}$, $\underline{z}$ its center. Thus $\underline{g} = \underline{k}' + \underline{z} + \underline{a} + \underline{n}$, and $U(\underline{g}) = (\underline{k}'U + U\underline{n}) + U(\underline{z})U(\underline{a})$. Let $u \to \emptyset_u$ be the projection: $U \to U(\underline{z})U(\underline{a})$ under this decomposition. Call $B = U(\underline{z})U(\underline{a})$, $\underline{c} = (\underline{z} + \underline{a})$. Note that $\underline{c}'$ (the dual of $\underline{c}$) $= \underline{z}' + \underline{a}'$.

Remark 10.1. If we give elements in B a commutative multiplication, then B is canonically isomorphic (as an algebra) to the polynomial algebra on $\underline{c}'$, which in turn is isomorphic to the symmetric algebra on $\underline{c}$.

The correspondence $T\colon B \to S(\underline{c})$ is as follows: let $v \in \underline{z}$, $w \in \underline{z}$, $\tau \in \underline{z}'$, $\lambda \in \underline{a}'$, $i, j \in Z^+$. Then $T(v^iw^j)(\tau + \lambda) = \langle \tau, v\rangle^i \langle \lambda, w\rangle^j$. Note that $p^\tau_{v^iw^j} = \langle \tau, v\rangle^i w^j$, and thus if $u \in U(\underline{z})U(\underline{a})$, $p^\tau_u(\lambda) = \emptyset_u(\tau + \lambda)$. Further, if $u \in (\underline{k}'U + U\underline{n})$, $p^\tau_u = \emptyset_u = 0$. Thus, if $u \in U$, $p^\tau_u(\lambda) = \emptyset_u(\tau + \lambda)$. Thus it suffices to evaluate the projection $\emptyset_u$.

Lemma 10.2. Define $D\colon S(\underline{c}) \to S(\underline{c}) \otimes S(\underline{c})$ as the unique algebra homomorphism sending X in $\underline{c}$ to $X \otimes 1 + 1 \otimes X$. Then if $r \in S(\underline{c})$, and $D(r) = \Sigma r_i' \otimes r_i''$, and $\lambda, \nu \in \underline{c}'$, then $r(\lambda + \nu) = \Sigma r_i'(\lambda) r_i''(\nu)$. Proof: Let $h\colon S(\underline{c}) \otimes S(\underline{c}) \to C$ be defined by $h(p \otimes q) = p(\lambda)q(\nu)$. Then $h \circ D$ is an algebra homomorphism: $S(\underline{c}) \to C$, agreeing on $\underline{c}$ with the homomorphism $f\colon r \to r(\lambda + \nu)$. Thus $h \circ D = f$.

Q.E.D.

We can now prove:

Proposition 10.3. Let $\lambda, \nu \in \underline{c}'$, $u \in U$, $D(u) = \Sigma u_i' \otimes u_i''$. Then $\phi_u(\lambda + \nu) = \Sigma\phi_{u_i'}(\lambda)\phi_{u_i''}(\nu)$. Proof: By Lemma 10.2, it suffices to prove that $D(\phi_u) = \Sigma\phi_{u_i'} \otimes \phi_{u_i''}$. So, call $U_0 = \underline{k}'U + U\underline{n}$. Then $U \otimes U = U_0 \otimes U_0 + B \otimes U_0 + U_0 \otimes B + B \otimes B$. Let P be the projection: $U \otimes U \to B \otimes B$ under this decomposition. Then note that $P(u \otimes v) = \phi_u \otimes \phi_v$. Now, $D(U_0) \subset U_0 \otimes U + U \otimes U_0$, so $P(D(u - \phi_u)) = 0$. Thus, $P(D(u)) = P(D(\phi_u))$. But $D(\phi_u) \in B \otimes B$ since $D(B) \subset B \otimes B$, so $P(D(\phi_u)) = D(\phi_u)$. Summarizing, $\Sigma\phi_{u_i'} \otimes \phi_{u_i''} = P(\Sigma u_i' \otimes u_i'') = P(D(u)) = D(\phi_u)$.

Q.E.D.

We now need a few definitions. Let $k \in Z^+$. Let D_k: $U \to U \otimes \cdots \otimes U$ (k times) be the map sending X in $\underline{g}$ to $(X \otimes 1 \otimes \cdots \otimes 1)+(1 \otimes X \otimes 1 \otimes \cdots \otimes 1)+ \cdots +(1 \otimes \cdots \otimes 1 \otimes X)$. Let $U_+ = U\underline{g}$, so that $U = U_+ + C\cdot 1$. Then $U \otimes \cdots \otimes U = (C\cdot 1 \otimes U \otimes \cdots \otimes U) + (U \otimes C\cdot 1 \otimes U \otimes \cdots \otimes U) + \cdots + (U \otimes \cdots \otimes U \otimes C\cdot 1) + U_+ \otimes \cdots \otimes U_+$.

Let $D_{k,+}$: $U \to U_+ \otimes \cdots \otimes U_+$, $D_{k,+}(u)$ = the component of $D_k(u)$ in $U_+ \otimes \cdots \otimes U_+$ under the decomposition above. Let $\nu \in \underline{c}'$. Define a linear map: $L_{\nu,k}$: $U \otimes \cdots \otimes U \to C$ by $L_{\nu,k}(u_1 \otimes \cdots \otimes u_k) = \prod_{i=1}^{k}\phi_{u_i}(\nu)$. Let $S_{\nu,k} = L_{\nu,k}\circ D_{k,+}$: $U \to C$. If $t \in C$, $m \in Z^+$, let $\binom{t}{m} = (m!)^{-1}\cdot t\cdot(t-1)\cdots(t-m+1)$. Let $U_{(n)}$ = elements in U of degree $\leq n$. We can now state:

Theorem 10.4. Let $u \in U(n)$, $\nu \in \underline{c}'$, $t \in C$. Then $\phi_u(t\nu) = \phi_u(0) + \sum_{m=1}^{n} S_{\nu,m}(u)\binom{t}{m}$. Proof: The proof is identical to that of Proposition 2.5.5 in ICSR, which gives the same result for p^0, so we will only outline a proof. First (compare with Lemma 10.2), one notes that if $p \in S(\underline{c})$, $\nu \in \underline{c}'$, then $p(k\nu) = L_{\nu,k}(D_k(p))$ $(k \in Z^+)$.

Second (compare with Proposition 10.3), one notes that if $u \in U$, and $D_k(u) = \sum_i u_i^{(1)} \otimes u_i^{(2)} \otimes \cdots \otimes u_i^{(k)}$ ($u_i^{(j)}$ in U), then $D_k(\phi_u) = \sum_i \phi_{u_i(1)} \otimes \cdots \otimes \phi_{u_i(k)}$. Combining, $\phi_u(k\nu) = L_{\nu,k}(D_k(u))$.

One then shows $L_{\nu,k}(D_k(u)) = \phi_u(0) + \sum_{0<m\leq k} S_{\nu,m}(u)\binom{k}{m}$. But one can easily see that if $k > n$, $S_{\nu,k}(u) = 0$. Thus $\phi_u(k\nu) = \phi_u(0) + \sum_{0<m\leq n} S_{\nu,m}(u)\binom{k}{m}$, for all k in Z^+. But $\phi_u(t\nu)$ is a polynomial in t, and $t \to \phi_u(0) + \sum_{m=1}^{n} S_{\nu,m}(u)\binom{t}{m}$ is also a polynomial in t, and the 2 polynomials agree at infinitely many distinct points (the non-negative integers). Thus, the 2 polynomials are the same, and for all t in C, $\phi_u(t\nu) = \phi_u(0) + \sum_{m=1}^{n} S_{\nu,m}(u)\binom{t}{m}$.

Q.E.D.

Section 11. Finite Dimensional Representations.

In this section we will obtain some information about the decomposition of a finite dimensional irreducible U-module into irreducible $\underline{k}$-submodules, for $\underline{g}_R = su(1,1)$ or $su(2,1)$. We will use this information to obtain some zeros in $\underline{c}'$ of ϕ_u, for various u in $S(\underline{d})^* = T^*$.

Let us first consider $\underline{g}_R = su(1,1) = sl(2R) = \{2 \times 2$ matrices with trace $0\}$. Then $\underline{k}_R = \{$skew-symmetric matrices in $\underline{g}_R\}$, $\underline{a}_R = \{$diagonal matrices in $\underline{g}_R\}$, and $\underline{n}_R = \{$strictly upper triangular matrices in $\underline{g}_R\}$. Let α be the unique element in Δ_+^1. Recall w in $\underline{a}$ is such that $\alpha(w) = 1$. Now,

$$\left[\begin{pmatrix}1 & 0\\ 0 & -1\end{pmatrix}, \begin{pmatrix}0 & x\\ 0 & 0\end{pmatrix}\right] = \begin{pmatrix}0 & 2x\\ 0 & 0\end{pmatrix}. \quad \text{Thus, } w = \begin{pmatrix}1 & 0\\ 0 & -1\end{pmatrix} \text{ and } \alpha\begin{pmatrix}a & 0\\ 0 & -a\end{pmatrix} = 2a,$$

for a, x in R. Also, $\theta\begin{pmatrix}0 & 1\\ 0 & 0\end{pmatrix} = \begin{pmatrix}0 & 0\\ -1 & 0\end{pmatrix}$, and since the Killing form in

$sl(n,R)$ is a scalar multiple of the form $(X,Y) = tr(XY)$, we let B_0 be the positive definite form $B_0(X,Y) = -tr(X\cdot\theta Y)$. Then $B_0(w) = \frac{1}{2}$, and

$\begin{pmatrix}0 & 1\\ 0 & 0\end{pmatrix} = 1$. Thus, recalling e_1 from section 9, $e_1 = \begin{pmatrix}0 & \frac{1}{2}\\ 0 & 0\end{pmatrix}$. Thus z = $i(e_1 - e_{-1}) = \frac{1}{2}i\begin{pmatrix}0 & 1\\ -1 & 0\end{pmatrix} \in \underline{k} = \underline{z}$, the center of $\underline{k}$.

Now, let $\tau_0 \in \underline{z}'$, be such that $\tau_0(z) = 1$. That is, $\tau_0\begin{pmatrix}0 & x\\ -x & 0\end{pmatrix} = -2ix$, for x in C. It is easily shown that τ_0 is the differential of the smallest character on $K \subset G = PSL(2,C)$. Also, τ_0 is twice the differential of the smallest character on $SO(2) \subset SL(2,R)$. We can now prove:

Proposition 11.1. Let V_λ be an irreducible finite dimensional U-module with highest weight $\lambda = n\alpha$. Then there exists a vector $v \in V_\lambda$ on which $\underline{k}$ acts according to $m\cdot\tau_0$ if and only if $m \in Z$, and $|m| \leq n$. Proof: We know that z is conjugate to -w. Thus, the eigenvalues of z on V_λ are those of -w, that is, $-n, -n+1, \ldots, n-1, n$. Thus, there exists v in V_λ such that $z\cdot v = mv$ if and only if $m \in Z$, $|m| \leq n$. Now, since $\tau_0(z) = 1$, and $\underline{k} = C\cdot z$, there exists v in V_λ such that $X\cdot v = m\cdot\tau_0(X)v$ for all X in $\underline{k}$ if and only if $m \in Z$, $|m| \leq n$.

Q.E.D.

Now we consider $\underline{g}_R = su(2,1) = \{$all 3 x 3 matrices of the form

$\begin{pmatrix}iX & w & v\\ -\overline{w} & iY & z\\ \overline{v} & \overline{z} & -i(X+Y)\end{pmatrix}$ X,Y in R, v,w,z in C$\}$. Then $\underline{k}_R = \left\{\begin{pmatrix}iX & w & 0\\ -\overline{w} & iY & 0\\ 0 & 0 & -i(Y+Y)\end{pmatrix} \text{ X,Y in R, w in C}\right\}$; $\underline{p}_R = \left\{\begin{pmatrix}0 & 0 & v\\ 0 & 0 & z\\ \overline{v} & \overline{z} & 0\end{pmatrix} \text{ v,z in C}\right\}$; $\underline{a}_R = \left\{\begin{pmatrix}0 & 0 & t\\ 0 & 0 & 0\\ t & 0 & 0\end{pmatrix} \text{ t in R}\right\}$; and

$\underline{n}_R = \underline{g}^{\alpha} + g^{2\alpha}$, where $\underline{g}^{2\alpha} = \left\{ \begin{pmatrix} -iX & 0 & iX \\ 0 & 0 & 0 \\ -iX & 0 & iX \end{pmatrix},\ X \in R \right\}$, $\underline{g}^{\alpha} = \left\{ \begin{pmatrix} 0 & c & 0 \\ -\bar{c} & 0 & \bar{c} \\ 0 & c & 0 \end{pmatrix},\ c \in C \right\}$.

Now, $w = \begin{pmatrix} 0 & 0 & 1 \\ 0 & 0 & 0 \\ 1 & 0 & 0 \end{pmatrix}$, and letting $B_0(X,Y) = -tr(X\cdot\theta Y)$, we get $B_0(w,w) = 2$, since $\theta = -1$ on $\underline{p}$, $+1$ on $\underline{k}$. Note that the Killing form is a scalar multiple of $(X,Y) = tr(X\cdot Y)$ for $su(2,1)$.

So, letting $e_1 = \frac{1}{2}\begin{pmatrix} 0 & 1 & 0 \\ -1 & 0 & 1 \\ 0 & 1 & 0 \end{pmatrix}$, we get $B_0(e_1,e_1) = 1$. So, $z = i(e_1 - e_{-1})$ $= i\begin{pmatrix} 0 & 1 & 0 \\ -1 & 0 & 0 \\ 0 & 0 & 0 \end{pmatrix} \in i\underline{k}_R$.

<u>Remark 11.2.</u> We have $z \in \underline{k}'$, the commutator subalgebra of $\underline{k}$.

Proof: $\left[\begin{pmatrix} 0 & i & 0 \\ i & 0 & 0 \\ 0 & 0 & 0 \end{pmatrix}, \begin{pmatrix} i & 0 & 0 \\ 0 & -i & 0 \\ 0 & 0 & 0 \end{pmatrix}\right] = \begin{pmatrix} 0 & 2 & 0 \\ -2 & 0 & 0 \\ 0 & 0 & 0 \end{pmatrix}$, and all 3 matrices lie in $\underline{k}_R$.

Q.E.D.

Now, if $\underline{z}$ is the center of $\underline{k}$, then $\underline{z} = C\cdot\begin{pmatrix} 1 & 0 & 0 \\ 0 & 1 & 0 \\ 0 & 0 & -2 \end{pmatrix}$. Let $\tau_0 \in \underline{z}'$ be such that $\tau_0 \begin{pmatrix} a & 0 & 0 \\ 0 & a & 0 \\ 0 & 0 & -2a \end{pmatrix} = -2a$. Then τ_0 can be extended to a character on $\underline{k}$. Also, τ_0 is the differential of the smallest character on K_0 $(\cong U(2))$, the maximal compact subgroup of $G_0 = SU(2,1)$. Note, however, that $3\tau_0$ is the differential of the smallest character on $K \subset G$, the complexified adjoint group of $su(2,1)$. For, τ_0 $(\text{on } U(2) \subset SU(2,1))$ sends $\begin{pmatrix} e^{i\theta} & 0 & 0 \\ 0 & e^{i\theta} & 0 \\ 0 & 0 & e^{i\theta} \end{pmatrix}$ to $e^{-2i\theta}$, and thus does not vanish on the center of $SU(2,1) = e^{(2\pi in/3)}\cdot \begin{pmatrix} 1 & 0 & 0 \\ 0 & 1 & 0 \\ 0 & 0 & 0 \end{pmatrix}$, $n \in Z$, while $\tau_0{}^3$ does vanish on the center and thus induces a character on K.

Now, note that $\underline{m}_R = \left\{ \begin{pmatrix} iX & 0 & 0 \\ 0 & -2iX & 0 \\ 0 & 0 & iX \end{pmatrix}, X \text{ in } R \right\}$, and thus $\tau|_{\underline{m}_R} \neq 0$ if $\tau \neq 0$.

In fact, a multiple of τ_0 is defined on all of $\underline{k}$ by its restriction to $\underline{m}$. Also, $\underline{h} = \underline{m} + \underline{a}$ is a Cartan subalgebra of $\underline{g}$, and any linear functional on $\underline{h}$ is of the form $m\cdot\tau_0|_{\underline{m}} + n\alpha$, m,n in C. From now on, we will write τ for $\tau|_{\underline{m}}$. We now prove:

<u>Theorem 11.3</u>. Let $\underline{g}_R = su(2,1)$, V_μ a U-module of highest weight μ, where $\mu|_{\underline{a}} = n\alpha$, $n \in Z^+$. Then V_μ has a vector v on which $\underline{k}$ acts according to $m\cdot\tau_0$ if and only if (a) $m \in Z$, (b) $\tau|_{\underline{m}} = m\cdot\tau_0$, (c) $m + n$ is even, and (d) $|m| \leq n$.

Proof: To prove this theorem, we must look carefully at the possible dominant integral weights (and corresponding irreducible finite dimensional representations) of $\underline{g} = sl(3,C)$.

First, we wish to find a set of positive roots such that $n\alpha$ is a dominant linear functional on $\underline{h}$ for $n \geq 0$. Now, $\begin{pmatrix} 0 & 1 & 0 \\ 0 & 0 & 0 \\ 0 & 1 & 0 \end{pmatrix}$, $\begin{pmatrix} 0 & 0 & 0 \\ 1 & 0 & 1 \\ 0 & 0 & 0 \end{pmatrix}$, and $\begin{pmatrix} -1 & 0 & 1 \\ 0 & 0 & 0 \\ -1 & 0 & 1 \end{pmatrix}$ are eigenvectors of $ad(\underline{h})$ corresponding to the roots $(\alpha + 3\tau_0)$, $(\alpha - 3\tau_0)$ and 2α respectively. We will set $(\alpha \pm 3\tau_0)$ as the 2 simple roots.

Note that $\underline{m}$ and $\underline{a}$ are orthogonal under B_0. Let $H_{\tau_0} \in \underline{m}$, $H_\alpha \in \underline{a}$, be such that, for all $X \in \underline{h}$, $B_0(H_{\tau_0},X) = \tau_0(X)$, $B_0(H_\alpha,X) = \alpha(X)$. Then we get $H_\alpha = \frac{1}{2}\begin{pmatrix} 0 & 0 & 1 \\ 0 & 0 & 0 \\ 1 & 0 & 0 \end{pmatrix}$, and $H_0 = (1/6)\begin{pmatrix} -1 & 0 & 0 \\ 0 & 2 & 0 \\ 0 & 0 & -1 \end{pmatrix}$. Thus, extending B_0 to $\underline{h}'$, we get $B_0(\alpha,\alpha) = \frac{1}{2}$, and $B_0(\tau_0,\tau_0) = 1/6$. Thus, since $B_0(\tau_0,\alpha) = 0$, $B_0(\alpha \pm 3\tau_0, \alpha \pm \tau_0) = 0$, and $B_0(\alpha \pm 3\tau_0, \alpha \pm \tau_0) = 1$, that is, the fundamental weights are $(\alpha \pm \tau_0)$.

Now we can prove the theorem in one direction. Suppose V_μ is a finite dimensional irreducible U-module with a vector v on which $\underline{k}$ acts according to $m\cdot\tau_0$. Now, let v_+ be the highest weight vector on V_μ. So, $V_\mu = U(\underline{g})\cdot v_+ = U(\underline{k})U(\underline{a})U(\underline{n})\cdot v_+$. But it is easy to see that $U(\underline{a})U(\underline{n})\cdot v_+ = C\cdot v_+$, so $U(\underline{k})\cdot v_+ = V_\mu$.

Now, V_μ is completely reducible as a $\underline{k}$-module, so let $V_\mu = V_0 + V_1 + \ldots + V_r$ as a sum of irreducible $\underline{k}$-modules, where $V_0 = C\cdot v$. Let P_o: $V_\mu \to V_0$ be the ($\underline{k}$-linear) projection under this decomposition. Then $P_0(v_+) \neq 0$ since $U(\underline{k})\cdot v_+ = V_\mu$. Further, P_0 is $\underline{m}$-linear, and since $C\cdot v_+$ and V_0 are 1-dimensional and $\underline{m}$-stable, P_o: $C\cdot v_+ \to V_0$ is an isomorphism of $\underline{m}$-modules. Thus, the action of $\underline{m}$ on v_+ is the same as it is on v, that is, $m\cdot\tau_0$. But the action of

$\underline{m}$ on v_+ is $\mu|_{\underline{m}}$, and thus $\mu|_{\underline{m}} = m\cdot\tau_0$, proving (b).

Now, calling $\lambda_1 = (\alpha + \tau_0)$, $\lambda_2 = (\alpha - \tau_0)$, the dominant integral weights of $\underline{g}$ are $\{r\lambda_1 + s\lambda_2,\ r,s \text{ in } Z^+\}$. Since $\mu|_{\underline{a}} = n\alpha$, and $\mu|_{\underline{m}} = m\cdot\tau_0$, $\mu = m\tau_0 + n\alpha = r\lambda_1 + s\lambda_2$, where $n = r + s$, $m = r - s$. Thus, $m \in Z$, proving (a), $|m| \leq n$, proving (d), and $m + n = 2r$ is even, proving (c).

For the converse, given (a) through (d), suppose we can prove that there exists a $\underline{k}$-stabilized 1-dimensional subspace V_0 of V_μ. Then again, the $\underline{k}$-linear projection P_0: $V \to V_0$ is non-0 on v_+. Again, P_0 is an $\underline{m}$-linear isomorphism: $C\cdot v_+ \to V_0$, so the action of $\underline{m}$ on V_0 is $\mu|_{\underline{m}} = m\cdot\tau_0$. But the commutator $\underline{k}'$ of $\underline{k}$ is semisimple, and thus $\underline{k}'\cdot V_0 = 0$. Since $\underline{m}$ and $\underline{k}'$ span $\underline{k}$, we see that the action of $\underline{k}$ on V_0 is $m\cdot\tau_0$. Thus, it suffices to prove that there exists a $\underline{k}$-stabilized 1-dimensional subspace, that is, that there exist a $\underline{k}'$-annihilated vector.

Now, consider the action of ad z on $\underline{g}$. Since z is conjugate to -w, there are 2 eigenvectors corresponding to the eigenvalue 0, 2 corresponding to +1, two to -1, one to +2, and one to -2. Now, $\underline{k}$ and $\underline{p}$ are ad($\underline{k}$)-modules, so each eigenvector must lie in $\underline{k}$ or $\underline{p}$. The vectors corresponding to 0 are z and an element of $\underline{z}$. The vectors corresponding to +1 are g and h (see section 9), lying in $\underline{d} \subset \underline{p}$. The vectors corresponding to -1 are $\bar{g}$, $\bar{h} \in \underline{p}$. Thus we see that there are elements $X^\pm \in \underline{k}'$ corresponding to the eigenvalues ± 2. Note that $X^\pm \notin \underline{k}_R{}'$.

Now, recall $c \in G$ such that $c\cdot(-w) = z$. Let $c^{-1}\cdot X^\pm = Y^\pm$. Then $Y^\pm \in \underline{g}^{\pm 2\alpha}$. Thus, since $(-w, Y^+, Y^-)$ span a copy $\underline{s}$ of sl(2,C) in $\underline{g}$, (z, X^+, X^-) also span a copy $\underline{t}$ of sl(2,C) in $\underline{g}$. Further, if $V_\mu = V_1 + \ldots + V_r$ as a direct sum of irreducible $\underline{s}$-modules, then $V_\mu = c^{-1}\cdot V_1 + \ldots + c^{-1}\cdot V_r$ as a sum of irreducible $\underline{t}$-modules.

Recall that it suffices to show (since $\underline{t} = \underline{k}'$), to prove the theorem, that there exists a $\underline{t}$-annihilated vector. Thus it suffices to show that there exists an $\underline{s}$-annihilated vector. One shows that there exists an $\underline{s}$-annihilated vector by carefully examining the possible sets of weights which can occur in an irreducible finite-dimensional sl(3,C)-module. One clearly wishes to find a vector v of weight σ such that $\sigma|_{\underline{a}} = 0$ (thus $-w\cdot v = 0$), $Y^+\cdot v = 0$, and $Y^-\cdot v = 0$. Now, according to (3a), p. 114, we have:

Lemma 11.4. Let V_μ be an sl(3,C)-module of highest weight μ. Then if λ is an integral linear combination of λ_1 and λ_2 (that is, an integral linear functional), then λ is a weight in V_μ if and only if, for all s in W, $\mu - s\cdot\lambda$ is a positive integral linear combination of simple roots. If λ is a weight, ν a root, then there exist integers r, q such that $(\lambda - j\nu)$ is a weight if and only if $j \in Z$, $-q \leq j \leq r$; further, $r - q = 2B_0(\lambda,\nu)/B_0(\nu,\nu)$.

It follows (see 3a.) that the weights which occur lie on a set of shells, the outer one of which is a hexagon or a triangle (here, a single point will be considered a triangle). Some possible shells are given in the diagram of $\underline{h}'$. The multiplicity of a weight (that is, the number of linearly independent vectors of that weight) on the outermost shell is 1, and the multiplicities increase by one each time one considers a shell one further inward, except that the multiplicities stabilize once the shell become triangles.

A Diagram of $\underline{h}'$.

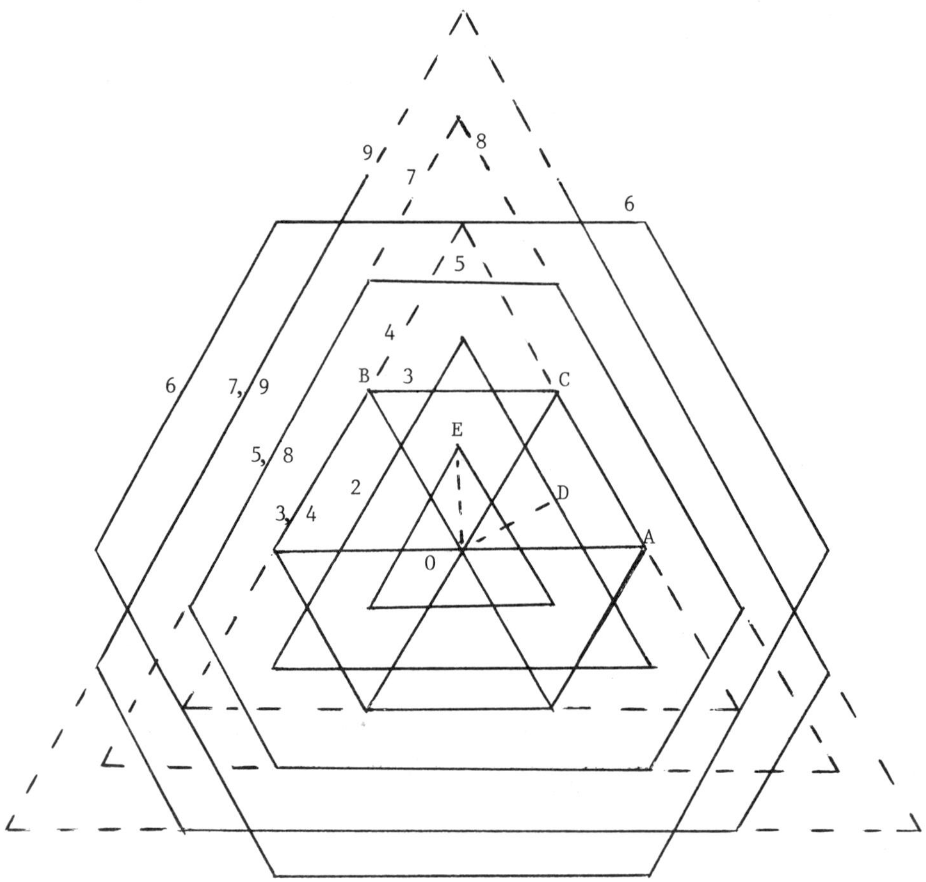

$A = \alpha - 3\tau_0$ $\quad$ $C = 2\alpha$ $\quad$ $E = \alpha + \tau_0$

$B = \alpha + 3\tau_0$ $\quad$ $D = \alpha - \tau_0$

In the following table, given that the highest weight is as specified in the first column, the weights which occur are those lying on the shells specified in the second column, with multiplicities in the 3rd column.

μ	shells	multiplicities	
λ_1	0	1	(trivial rep.)
$2\lambda_1$	2	1	(rep. on C^3)
$\lambda_1 + \lambda_2$	3, 0	1, 2	(adjoint rep.)
$2\lambda_1 + \lambda_2$	5, 1	1, 2	
$3\lambda_1$	4, 0	1, 1	
$2\lambda_1 + 2\lambda_2$	6, 3, 1	1, 2, 3	
$3\lambda_1 + \lambda_2$	7, 2	1, 2	
$4\lambda_1$	8, 1	1, 1	
$5\lambda_1$	9, 2	1, 1	

We can assume without loss of generality that $\mu = r\lambda_1 + s\lambda_2$, with $r \geq s$. Now we will prove the theorem in 2 separate cases: (a) All the shells are triangles (this occurs if $s = 0$), (b) the outermost shell is a hexagon (if $s > 0$).

In case (a), $\mu = r\lambda_1$. Now, consider the weight $-2r\tau_0 = r(\lambda_1 - \lambda_2)$. This corresponds to the lower right vertex of the outermost triangle. Now, $-2r\tau_0$ occurs with multiplicity 1, and $(-2r\tau_0 \pm 2\alpha)$ are not weights of V_μ. Thus, if v is a vector of weight $(-2r\tau_0)$, $w \cdot v = \tau(w)v = 0$, and $Y^{\pm} \cdot v$ have weight $(-2r\tau_0 \pm 2\alpha)$, and thus are 0. So, $\underline{s} \cdot v = 0$.

In case (b), the outer shell is a hexagon, so some shell is the outermost triangular shell (the diagram on p. 36 shows that the innermost shell is always triangular). Further, the weight corresponding to the lower right vertex of this shell is $(-2r\tau_0)$, for some r in Z^+, and this weight occurs with multiplicity $k > 1$. Now, $(-2r\tau_0 \pm 2\alpha)$ do occur as weights: they lie on the next outer shell (the innermost hexagonal shell), and thus have multiplicity $(k-1)$. Thus, letting W_λ be the weight space corresponding to a weight λ, $\dim W_{-2r\tau_0} = k$, and $Y^{\pm} \cdot W_{-2r\tau_0} \subset W_{(-2r\tau_0 \pm 2\alpha)}$, and thus has dimension $\leq (k-1)$. Thus, there exists v in $W_{-2r\tau_0}$ such that $Y \cdot v = 0$. Also, since $\tau_0(-w) = 0$, $-w \cdot v = 0$. So, consider the $U(\underline{s})$-module generated by v. It has highest weight vector v, and thus highest weight 0. Thus, by the representation theory of $sl(2,C)$, it is the 1-dimensional, trivial $U(\underline{s})$-module. That is, $\underline{s} \cdot v = 0$.

So, since either (a) or (b) occurs, we see that there exists an $\underline{s}$-annihilated vector, and thus there exists a $\underline{k}$-stable 1-dimensional subspace, and thus the action of $\underline{k}$ on this space is $m \cdot \tau_0$.

Q.E.D.

Proposition 11.1 and Theorem 11.3 are generalizations [for su(1,1) and su(2,1)] of a theorem of Helgason (2b.) on the necessary and sufficient conditions for a K-fixed vector in a finite-dimensional G-module. Using Proposition 11.1 and Theorem 11.3 we now prove:

Proposition 11.5.

(a) Leg $\underline{g}_R = su(1,1)$. Let $u \in T_j^* = \beta(T_j)$, $j \in Z^+$. Let $m \in Z$, $n \in Z^+$, $|m| \leq n$, $j - m > n$. Then $\emptyset_u(m\tau_0 + n\alpha) = 0$.

(b) Let $\underline{g}_R = su(2,1)$, $u \in T_j^*$, $j \in Z^+$. Let $m \in Z$, $n \in Z^+$, $\frac{1}{2}(m+n) \in Z$, $j > n \geq |m|$. Then $\emptyset_u(m\tau_0 + n\alpha) = 0$.

Proof:

(a) Let V be a finite dimensional irreducible U-module with highest weight $n\alpha$, and highest weight vector v^+. Let V' be the dual U-module. By Proposition 11.1, there exists a vector on which $\underline{k}$ acts according to $m\tau_0$; thus there exists $0 \neq x \in V'$ on which $\underline{k}$ acts according to $-m\tau_0$. Now, $U\cdot v^+ = V = U(\underline{k})U(\underline{a})U(\underline{n})v^+$. Now, $0 \neq \langle x,V\rangle = \langle x,U(\underline{k})U(\underline{a})U(\underline{n})v^+\rangle = \langle U(\underline{k})x,U(\underline{a})U(\underline{n})v^+\rangle = \langle C\cdot x,C\cdot v^+\rangle$. Thus, $\langle x,v^+\rangle \neq 0$, and so we can assume x is chosen so that $\langle x,v^+\rangle = 1$.

Now, recalling $\underline{k}^{\#}$, $\overline{\underline{k}^{\#}}$ from section 3 (chosen with respect to $\tau = m\tau_0$), we see that $\langle x,\underline{k}^{\#}U + U\underline{n})v^+\rangle = 0$, since $\underline{k}^{\#t} = \overline{\underline{k}^{\#}}$, and $\overline{\underline{k}^{\#}}\cdot x = 0 = \underline{n}\cdot v^+$. Thus, for u in U, $\langle x,u\cdot v^+\rangle = \langle x,p_u^{m\tau_0}\cdot v^+\rangle = p_u^{m\tau_0}(n\alpha) = \emptyset_u(m\tau_0 + n\alpha)$.

Now, since $u \in T_j^*$, $[z,u] = ju$. Since $z \in \underline{g}$, $[z,u]^t = [z,u^t] = ju^t$. Now, $z\cdot x = -m\tau_0(z)x = -mx$. Thus, $z\cdot u^t\cdot x = [z,u^t]x + u^t\cdot z\cdot x = ju^t\cdot x + u^t\cdot(-mx) = (j-m)u^t\cdot x$. So, $u^t x$ is an eigenvector for z in V' with eigenvalue $(j-m)$. But the eigenvalues of z are those of $-w$, and the highest eigenvalue of $-w$ in V' is n, and we have assumed that $j - m > n$. Thus, $u^t\cdot x = 0$. So, $0 = \langle u^t\cdot x,v^+\rangle = \langle x,u\cdot v^+\rangle = \emptyset_u(m\tau_0 + n\alpha)$.

(b) Here, let V be a finite dimensional irreducible U-module with highest weight $\mu = m\tau_0 + n\alpha$. By Theorem 11.3, V exists, and has a vector on which $\underline{k}$ acts according to $m\tau_0$.

Thus the dual module V' has a vector x on which $\underline{k}$ acts according to $-m\tau_0$, and, as in (a), x can be normalized so that, for all u in U, $\langle x,u\cdot v^+\rangle = p_u^{m\tau_0}(n\alpha) = \emptyset_u(m\tau_0 + n\alpha)$. Again, $[z,u^t] = ju^t$. But, by Remark 11.2, $z \in \underline{k}'$, and so $\tau_0(z) = 0$. Thus, $z\cdot x = 0$, and $[z,u^t]\cdot x = z\cdot u^t\cdot x = ju^t\cdot x$. Thus $u^t x$ is an eigenvector for z with eigenvalue j, but, as in (a), the highest eigenvalue for $-w$ and thus for z in V' is n, and $j > n$, so $u^t\cdot x = 0$. Thus, $0 = \langle u^t\cdot x,v^+\rangle = \emptyset_u(m\tau_0 + n\alpha)$.

Q.E.D.

Section 12. Calculating p_γ^τ for su(N,1).

In this section, we compute p_γ^τ as γ ranges through Γ for su(N,1). We first consider the case $N \geq 2$. By Corollary 9.7, we have reduced the problem to finding $\{p^\tau_{(g + ih)^j(g - ih)^k}, j,k \in Z^+\}$. Section 2.8 of ICSR gives us the following result:

Lemma 12.1. $p^0_{(g + ih)^j(g - ih)^k} = 2^{-j-k} \cdot w(w - 2)\cdots(w - 2j + 2)w(w-2)\cdots(w - 2k + 2)$.

Recalling the projection $\emptyset$ of section 10, and that $p_u^\tau(\lambda) = \emptyset_u(\tau + \lambda)$, we compute $\emptyset_{(g+ih)^j(g-ih)^k}$.

Let us review from Section 2.7 of ICSR how g and h are found. We pick e_1 in $\underline{g}^\alpha$, such that $B_0(e_1,e_1) = \frac{1}{2}B_0(w,w)$. Then $g = \frac{1}{2}(w - i(e_1 + e_{-1}))$. Also, let $q \in \underline{g}^\alpha$, $B_0(e_1,q) = 0$, $B_0(q,q) = (1/8)B_0(w,w)$. Then $h = q + \theta q + i[e_1,q] + i\theta[e_1,q]$. Call $y = [e_1,q]$. Then Section 2.7 of ICSR shows $B_0(y,y) = \frac{1}{4}B_0(w,w)$. Now, e_1, q, iy, e_{-1}, θq, $i\theta y$, w, and $[e_{-1},q]$ span a copy of su(2,1) in su(N,1). This can be shown by explicit calculation.

Let us write su(N,1) as the set of (N+1) x (N+1) matrices of the form $\begin{pmatrix} A & B \\ \overline{B}^t & c \end{pmatrix}$, where A is an N x N block, such that $\overline{A}^t = -A$, B an N x 1 block of complex numbers, c in iR. Matrices of this form with B = 0 lie in $\underline{k}_R$, those with A = 0, c = 0 lie in $\underline{p}_R$. Choose $\underline{a}_R$ to be all real multiples of w =

$$\begin{pmatrix} 0 & 0 & \cdots & 0 & 1 \\ \cdot & & & \cdot & \\ \cdot & & & \cdot & \\ \cdot & & & \cdot & \\ 0 & & \cdots & 0 & 0 \\ 1 & 0 & \cdots & 0 & 0 \end{pmatrix} : \text{ then } e_1 = \tfrac{1}{2}\begin{pmatrix} 0 & 1 & 0 & \cdots & 0 \\ -1 & 0 & \cdots & 0 & 1 \\ \cdots & \cdots & \cdots & \cdots & \cdots \\ 0 & 1 & 0 & \cdots & 0 \end{pmatrix} \in \underline{g}^\alpha,$$

$$q = \tfrac{1}{4}\begin{pmatrix} 0 & i & \cdots & & 0 \\ 1 & 0 & \cdots & 0 & i \\ \cdots & \cdots & \cdots & \cdots & \cdots \\ 0 & -i & 0 & \cdots & 0 \end{pmatrix} \in \underline{g}^\alpha, \quad iy = \tfrac{1}{4}i\begin{pmatrix} -1 & 0 & \cdots & 0 & 1 \\ 0 & 0 & \cdots & 0 & 0 \\ \cdots & \cdots & \cdots & \cdots & \cdots \\ -1 & 0 & \cdots & 0 & 1 \end{pmatrix} \in \underline{g}^{2\alpha}.$$ Then we easily

see that the elements above generate a copy of su(2,1) consisting of:

$$\begin{pmatrix} iY & X & 0 & \cdots & 0 & W \\ -\overline{X} & iZ & 0 & \cdots & 0 & V \\ \cdots & \cdots & \cdots & \cdots & \cdots & \cdots \\ W & \overline{V} & 0 & \cdots & 0 & -iY-iZ \end{pmatrix}.$$

Further, if we let $\underline{z}_0$ be the center of $\underline{k}$ for this copy of su(2,1), then $\underline{g} = \underline{k}' + \underline{z}_0 + \underline{a} + \underline{n}$, and thus $U = (\underline{k}'U + U\underline{n}) + U(\underline{z}_0)U(\underline{a})$, and if $u \to \emptyset_u'$ is the projection to $U(\underline{z}_0)U(\underline{a})$ under this decomposition, it is easy to see that we still have $p_u^\tau = p^\tau_{\emptyset_u'}$. Thus we will actually calculate $\emptyset_u'$ for u in T^*. From now on we will just call this projection $\emptyset$.

Recall τ_0 is the differential of the smallest character on $K_0 = U(N) \subset SU(N,1) = G_0$. Call v the element of $\underline{z}_0$ such that $\tau_0(v) = 1$. Then v =

$\begin{pmatrix} -\frac{1}{2} & 0 & \cdots & 0 & 0 \\ 0 & -\frac{1}{2} & 0 & \cdots & 0 \\ \cdots & \cdots & \cdots & \cdots & \cdots \\ 0 & \cdots & \cdots & 0 & 1 \end{pmatrix}$. But we also have $\tau_0(v) = 1$ if we look at v as it sits inside su(2,1): $v = -\begin{pmatrix} \frac{1}{2} & 0 & 0 \\ 0 & \frac{1}{2} & 0 \\ 0 & 0 & -1 \end{pmatrix}$. So, from now on we need only consider the algebra su(2,1).

Calculating, we get: $y = \frac{1}{4}\begin{pmatrix} -1 & 0 & 1 \\ 0 & 0 & 0 \\ -1 & 0 & 1 \end{pmatrix}$, and $(g \pm ih) = \begin{pmatrix} \mp\frac{1}{2}i & \pm\frac{1}{2}(i-1) & \frac{1}{2} \\ \mp\frac{1}{2}(i-1) & 0 & i \\ \frac{1}{2} & i & \pm\frac{1}{2}i \end{pmatrix} = \frac{1}{2}(w \mp v) + X + Y$, where $X \in \underline{k}'$, $Y \in \underline{n}$. Thus, $\emptyset_{(g \pm ih)} = \frac{1}{2}(w \mp v)$. Let $u_\pm = (g \pm ih)$. Now we start to prove:

Theorem 12.2. $\emptyset_{u_+^j u_-^k} = 2^{-j-k}(w-v)(w-v-2)\cdots(w-v-2j+2) \cdot (w+v)(w+v-2)\cdots(w+v-2k+2)$. We start with:

Proposition 12.3. Let $k \in Z^+$. Then $\emptyset_{u_\pm^k} = c\cdot(w \mp v)\cdots(w \mp v-2k+2)$, where c is a scalar. Proof: We first prove:

Lemma 12.4. $\emptyset_{u_\pm^k}\big(t(\alpha \pm \tau_0)\big) = 0$, for all $t \in C$, $k \geq 1$. Proof: By Theorem 10.4, $\emptyset_{u_\pm^k}\big(t(\alpha \pm \tau_0)\big) = \emptyset_{u_\pm^k}(0) + \sum_{m=1}^{k} S_{(\alpha \pm \tau_0),m}(u_\pm^k)\binom{t}{m}$. But terms of $D_{k,+}(u_\pm^k)$ are of the form $c\cdot u_\pm^{r_1} \otimes \cdots \otimes u_\pm^{r_s}$, where $r_1 + \cdots + r_s = k$, and each $r_j \geq 1$. But $\emptyset_{u_\pm^{r_j}}(\alpha \pm \tau_0) = 0$ by Proposition 11.5, if $r_j > 1$. Also by Proposition 11.5, $\emptyset_{u_\pm^k}(0) = 0$. Thus we are left only with the term $c\cdot u_\pm \otimes \cdots \otimes u_\pm$, since in all the other cases some $u_\pm$ has exponent > 1. But $\emptyset_{u_\pm}(\alpha \pm \tau_0) = 0$ by direct computation, since $\emptyset_{u_\pm} = \frac{1}{2}(w \mp v)$. So, all the terms are 0, and $\emptyset_{u_\pm^k}\big(t(\alpha \pm \tau_0)\big) = 0$.

Q.E.D.

The following lemma implies Proposition 12.3.

Lemma 12.5. If $j > n$, $\emptyset_{u_\pm^j}\big(t(\alpha \pm \tau_0)+2n\alpha\big) = 0$. Proof: We apply Proposition 10.3 to $\lambda = t(\alpha \pm \tau_0)$, $\nu = 2n\alpha$. $D(u_\pm^j) = u_\pm^j \otimes 1 + j\cdot u_\pm^{j-1} \otimes u_\pm + \cdots + j\cdot u_\pm \otimes u_\pm^{j-1} + 1 \otimes u_\pm^j$. Thus, $\emptyset_{u_\pm^j}\big(t(\alpha \pm \tau_0)+2n\alpha\big) = \emptyset_{u_\pm^j}\big(t(\alpha \pm \tau_0)\big)\emptyset_1(2n\alpha) + \cdots + j\cdot\emptyset_{u_\pm}\big(t(\alpha \pm \tau_0)\big)\emptyset_{u_\pm^{j-1}}(2n\alpha) + 1\cdot\emptyset_{u_\pm^j}(2n\alpha)$. Now, the left hand factor in each term in the above expression except for the last term is 0 by Lemma 12.4. But, looking at the last term, $\emptyset_{u_\pm^j}(2n\alpha) = p^0_{u_\pm^j}(2n\alpha) = 0$ by Lemma 12.1. Thus $\emptyset_{u_\pm^j}\big(t(\alpha \pm \tau_0)+2n\alpha\big) = 0$.

Q.E.D.

So, we have found that $(w \mp v-2n)$ is a factor of $\emptyset_{u_\pm^j}$ for $0 \leq n < j$, since $\emptyset_{u_\pm^j}$ vanishes on the line $\big(t(\alpha \pm \tau_0)+n\alpha\big)$, t in C. So, we have found j linear factors for a polynomial of degree j, thus proving the proposition.

Q.E.D.

Now we wish to prove the theorem. First, we prove that, up to scalar, $\emptyset_{u_+^j u_-^k} = (w-v)\cdots(w-v-2j+2)\ (w+v)\cdots(w+v-2k+2)$. We know this result if either $j = 0$ or $k = 0$. We prove the result by induction: we assume we know it for (J,K): $0 \leq J < j$, $0 \leq K \leq k$, and for $0 \leq J \leq j$, $0 \leq K < k$, and then we prove it for (j,k).

First note that $\emptyset_{u_+^j u_-^k}(t(\alpha \pm \tau_0)) = 0$. For, every term of $D_{m,+}(u_+^j u_-^k)$ $(m \leq j + k)$ has a factor $u_+^J u_-^K$ where $0 \leq J < j$, and $0 < K < k$, and a factor (maybe the same) $u_+^J u_-^K$, for $0 < J < j$, and $0 \leq K < k$. In the former case $\emptyset_{u_+^J u_-^K}(t(\alpha - \tau_0)) = 0$, and in the latter case $\emptyset_{u_+^J u_-^K}(t(\alpha + \tau_0)) = 0$, by the induction hypothesis (since $(w \mp v)$ is a factor of $\emptyset_{u_+^J u_-^K}$). So, every term is 0 by Theorem 10.4, since $\emptyset_{u_+^j u_-^k}(0) = 0$ by Proposition 11.5.

Thus, $(w \mp v)$ is a factor of $\emptyset_{u_+^j u_-^k}$ if $j \geq 1$, $k \geq 1$. Now we show, using Proposition 10.3, that if $j > n$, $\emptyset_{u_+^j u_-^k}(t(\alpha + \tau_0) + 2n\alpha) = 0$ for t in C. The identical argument shows that if $k > n$, $\emptyset_{u_+^j u_-^k}(t(\alpha - \tau_0) + 2n\alpha) = 0$.

Now, $D(u_+^j u_-^k)$ consists of terms of the type (1) $u_+^J u_-^K \otimes u_+^{j-J} u_-^{k-K}$, $0 < J \leq j$, $0 \leq K \leq k$, and terms (2) $u_-^K \otimes u_+^j u_-^{k-K}$, $0 \leq K \leq k$. But $\emptyset_{u_+^J u_-^K}(t(\alpha + \tau_0)) = 0$ if $0 < J \leq j$ and $0 \leq K \leq k$, since $(w-v)$ is a factor of $\emptyset_{u_+^J u_-^K}$ by the previous discussion. Thus terms of the first type contribute 0. But by Lemma 12.1, $p^0_{u_+^j}(2n\alpha) = \emptyset_{u_+^j}(2n\alpha) = 0$ if $j > n$. Thus terms of the second type give a contribution of 0 also. Thus, $(w - v - 2n)$ is a (linear) factor of $\emptyset_{u_+^j u_-^k}$ if $j > n$. Similarly, we show that $(w + v + 2n)$ is a factor of $\emptyset_{u_+^j u_-^k}$ if $k > n$. Since we have exhibited $(j+k)$ factors of this polynomial of degree $(j+k)$, we have that up to scalar multiple, $\emptyset_{u_+^j u_-^k} = (w - v)(w - v - 2)\cdots(w-v-2j+2)\cdot(w+v)(w+v-2)\cdots(w+v=2k+2)$. But from Lemma 12.1, we see that the scalar multiple is 2^{-j-k}.

Q.E.D.

Now we consider the case $\underline{g}_R = su(1,1)$. Recall that τ_0 is the differential of the smallest character on $K \subset G = PSL(2,C)$, and $z = i(e_1 - e_{-1})$ is such that $\tau_0(z) = 1$. We write v for z. Note that $C \cdot v = \underline{k}$.

By Section 2.7 of ICSR, $g\ (\varepsilon\ \underline{d}) = \frac{1}{2}(w + i(e_1 - e_{-1})) = \frac{1}{2}(w + v) - ie_1$, and so $\emptyset_g = \frac{1}{2}(w + v)$. Recall (Corollary 9.7) that we are interested in calculating $\emptyset_{\bar{g}^s}$ and $\emptyset_{g^s}$, for s in Z^+. The calculation of the latter is identical to that of the former, so we will only calculate the latter. We have

<u>Theorem 12.6.</u> $\emptyset_{g^s} = 2^{-s}(w+v)(w+v-1)\cdots(w+v-s+1)$, and $\emptyset_{\bar{g}^s} = 2^{-s}(w-v)\cdot(w-v-1)\cdots(w-v-s+1)$.

We first prove:

<u>Lemma 12.7.</u> $\emptyset_{g^s}(t(\alpha - \tau_0)) = 0$ for all t in C, $s \geq 1$. Proof: It is

clearly sufficient to prove this for $t \in Z^+$, since a polynomial in 1 variable is determined by its values on the positive integers. We know the result for $s = 1$, so assume we know $\phi_{g^S}(N(\alpha-\tau_0)) = 0$, (a) for all $S < s$, $N \in Z^+$, and (b) for $S = s$, $N < n$. Then we wish to show $\phi_{g^s}(n(\alpha-\tau_0)) = 0$.

Now, $n(\alpha-\tau_0) = (n-1)(\alpha-\tau_0) + (\alpha-\tau_0)$. Applying Proposition 10.3, $D(g^s) = 1 \otimes g^s + g^s \otimes 1 +$ (terms with g^i in each factor, $0 < i < s$). And, we know that $\phi_{g^i}(\alpha-\tau_0) = 0$ for $0 < i \leqslant s$. So, we are left with only the term $\phi_{g^s}((n-1)(\alpha-\tau_0))\cdot\phi_1(\alpha-\tau_0)$. But by the induction assumption, $\phi_{g^s}((n-1)(\alpha-\tau_0)) = 0$. Thus, $\phi_{g^s}(n(\alpha-\tau_0)) = 0$ for all n in Z^+, and thus $\phi_{g^s}(t(\alpha-\tau_0)) = 0$ for all t in C.

Q.E.D.

<u>Lemma 12.8</u>. $\phi_{g^s}(t(\alpha+\tau_0)+n\alpha) = 0$ if $t \in C$, $n \leqslant s-1$. Proof: Again, we use Proposition 10.3. $D(g^s) = g^s \otimes 1 + \binom{s}{1}g^{s-1} \otimes g + \cdots + \binom{s}{1}g \otimes g^{s-1} + 1 \otimes g^s$. Now, $\phi_{g^j}(t(\alpha+\tau_0)) = 0$ if $j \geqslant 1$ (by the previous lemma), and by Theorem 2.7.11 of ICSR, $\phi_{g^s}(n\alpha) = p^0_{g^s}(n\alpha) = 0$ if $n \leqslant s-1$. Thus, using Proposition 10.3, we see that all the terms vanish, and that $\phi_{g^s}(t(\alpha+\tau_0)+n\alpha) = 0$ if $n \leqslant s-1$.

Q.E.D.

We, we have exhibited s distinct linear factors of ϕ_{g^s}, a polynomial of degree s, and thus up to scalar, $\phi_{g_s} = (w+v)(w+v-1)\cdots(w+v-s+1)$. To determine the scalar, note from Theorem 2.7.11 of ICSR that $\phi_{g^s}(t\alpha) = 2^{-s}\, t(t-1)\cdots(t-s+1)$.

In the same manner that we found ϕ_{g^s}, we can find that $\phi_{\overline{g^s}} = 2^{-s}(w-v)(w-v-1)\cdots(w-v-s+1)$.

Q.E.D.

<u>Corollary 12.9</u>.

(a) If $\underline{g}_R = su(1,1)$, then as γ runs through Γ, $p_\gamma^{m\tau_0}$ runs through the set of polynomials $\{(w\pm m)(w\pm m-1)\cdots(w\pm m-s+1),\ s \text{ in } Z^+\}$, and

(b) If $\underline{g}_R = su(N,1)$, $N \geqslant 2$, then as γ runs through Γ, $p_\gamma^{m\tau_0}$ runs through the set $\{(w\pm m)(w\pm m-2)\cdots(w\pm m-2j+2)(w\pm m)(w\pm m-2)\cdots(w\pm m-2k+2),\ j, k \in Z^+\}$.

Proof: This follows from Theorem 12.2, Theorem 12.6, Corollary 9.7, and the fact that $p_u^\tau(\lambda) = \phi_u(\tau+\lambda)$ for u in U, λ in $\underline{a}'$.

Chapter IV

Section 13. The Zeros of R_τ^γ and P_τ^γ.

Recall from Section 5 that $p_\tau^u(\lambda) = p_{u^t}^{-\tau}(\lambda')$. And, since $\kappa\theta$ stabilizes $\underline{k}^{\#}U$ and $U\underline{n}$, $p_{\kappa\theta\cdot u}^\tau = \kappa\theta\cdot p_u^\tau$. But if $p \in U(\underline{a})$, $p(\lambda') = (\kappa\theta\cdot p)(\lambda)$, since $\theta = -1$ on $\underline{p}$. Thus, $p_u^\tau(\lambda') = \kappa\theta\cdot p_u^\tau(\lambda) = p_{\kappa\theta\cdot u}^\tau(\lambda)$. Now, if $u \in S(\underline{p})$, $\theta\cdot u = u^t$, and thus $p_\tau^u(\lambda) = p_{u^t}^{-\tau}(\lambda') = p_{\kappa\theta\cdot u^t}^{-\tau}(\lambda) = p_{\kappa\cdot u}^{-\tau}(\lambda)$.

Now, let $\{\varepsilon_1,\ldots,\varepsilon_{1(\gamma)}\}$ be a basis for E_γ, $\{v_1,\ldots,v_{1(\gamma)}\}$ a basis for $V_\gamma^{M_\theta}$. Then $p_\tau^{\varepsilon_i(v_j)} = p_{\kappa\cdot\varepsilon_i(v_j)}^{-\tau} =$ (since ε_i is K_θ-linear) $p_{\varepsilon_i(\kappa\cdot v_j)}^{-\tau}$. But κ normalizes M_θ, so $\kappa\cdot V^{M_\theta} = V^{M_\theta}$, and therefore $\det\left(p_\tau^{\varepsilon_i(v_j)}\right) = \det\left(p_{\varepsilon_i(\kappa\cdot v_j)}^{-\tau}\right) = \det\left(p_{\varepsilon_i(v_j)}^{-\tau}\right)\det(\kappa|_{V_\gamma M_\theta})$, that is, $p_\tau^\gamma = c_\gamma p_\gamma^{-\tau}$, where $0 \neq c_\gamma$ is a constant.

Now recall (Theorem 7.1) that $R_\tau^\gamma(\lambda) = P_\tau^\gamma(\lambda)P_\tau^\gamma(-\bar{\lambda}-2\rho)^*$. So, $r_\tau^\gamma(\lambda) = p_\tau^\gamma(\lambda)\overline{p_\tau^\gamma(-\bar{\lambda}-2\rho)} = |c_\gamma|^2 p_\gamma^{-\tau}(\lambda)\overline{p_\gamma^{-\tau}(-\bar{\lambda}-2\rho)} = |c_\gamma|^2 p_\gamma^{-\tau}(\lambda)\overline{p_\gamma^{-\tau}}(-\lambda-2\rho)$. But our expressions for su(n,1) in section 12, and the expressions for p_γ^0 in Corollary 2.4.8 of ICSR, along with Theorem 8.1, show that $\overline{p_\gamma^{-\tau}} = p_\gamma^\tau$. So, up to positive real scalar, $r_{-\tau}^\gamma(\lambda) = p_\gamma^\tau(\lambda)p_\gamma^\tau(-\lambda-2\rho)$, and up to scalar, $p_\tau^\gamma(\lambda) = p_\gamma^{-\tau}(\lambda)$.

Now, to find when $p_\gamma^\tau(\lambda) = 0$ for some $\gamma \in \Gamma$, one must find, by section 8, when $p_{\delta_i\cdot\phi}^\tau(\lambda) = 0$ for some $\emptyset$ in Δ_+^1, $\delta_i\cdot\phi$ as defined in section 8. But it is easy to see that, as γ ranges through Γ, the δ^ϕ range through all of $\Gamma_\emptyset$. For H in $\underline{a}$, let $\rho_\emptyset(H) = \frac{1}{2}\mathrm{tr}\ \mathrm{ad}\ H|_{\underline{n}_\emptyset}$. Thus, since p_γ^τ is a product of elements $T_{\rho_\phi-\rho}P_{\delta\phi}$, we see $p_\gamma^\tau(\lambda) = 0$ for some $\gamma \in \Gamma$ if and only if there exists $\emptyset$ in Δ_+^1, $\delta^\emptyset$ in $\Gamma_\emptyset$, such that $T_{\rho_\phi}p_{\delta\phi}^\tau(\lambda+\rho) = 0$.

Now, recall $p_{\delta\phi}^\tau$ is a polynomial in $w_\emptyset \in [\underline{g}_\emptyset,\underline{g}_\emptyset] \cap \underline{a}$, where $\emptyset(w_\emptyset) = 1$. Let $s_\emptyset = \dim \underline{g}^{2\emptyset}$, $k_\emptyset = s_\emptyset + \frac{1}{2}\dim \underline{g}^\emptyset$. Then $\rho_\emptyset(w_\emptyset) = k_\emptyset$. So $T_{\rho_\emptyset}(w_\emptyset) = w_\emptyset - k_\emptyset$.

Now, in Section 2.9 of ICSR, it is noted that as δ^ϕ runs through $\Gamma_\emptyset$, $T_{\rho_\phi}P_{\delta\phi}^0$ runs through the set of polynomials $(w_\emptyset-k_\emptyset)(w_\emptyset-k_\emptyset-2)\cdots\left(w_\emptyset-k_\emptyset-2(i+j)+2\right)\cdot(w_\emptyset-k_\emptyset-1+s_\emptyset)(w_\emptyset-k_\emptyset-3+s_\emptyset)\cdots(w_\emptyset-k_\emptyset-2i+1+s_\emptyset)$, where $i,j \in Z^+$ if $s_\emptyset > 0$, $\cdot$ $i \in Z^+$, $j = 0$ or 1 if $s_\emptyset = 0$.

Now suppose τ is a character on $\underline{k}$, $\tau|_{\underline{k}_\emptyset} \neq 0$. We have seen that this implies $(\underline{g}_\emptyset)_R = su(N,1)$, $N \geq 1$. If $(\underline{g}_\emptyset)_R = su(1,1)$, let $\tau_{\emptyset,0}$ be the character on $\underline{k}_\emptyset$ which is the differential of the smallest character on $k_\emptyset \subset G_\emptyset \cong PSL(2,C)$. If $(\underline{g}_\emptyset)_R = su(N,1)$, $N \geq 2$, let $\tau_{\emptyset,0}$ be the character on $\underline{k}_\emptyset$ which is differential of the smallest character on K_0 $\left(\cong U(N)\right) \subset G_0 = SU(N,1)$. Let $m_\emptyset$ be such that $\tau|_{\underline{k}_\emptyset} = m_{\emptyset,0}$. Then $m_\emptyset$ is defined by τ up to sign. Let $m_\emptyset = 0$ if $\tau|_{\underline{k}_\emptyset} = 0$.

Now, combining Corollary 2.4.8 of ICSR and Corollary 12.9, it is easy to see that as δ runs through $\Gamma_\emptyset$, $T_{\rho\emptyset}p^\tau_\delta$ runs through elements of the form $(w_\emptyset - k_\emptyset \pm m_\emptyset)(w_\emptyset - k_\emptyset - 2 \pm m_\emptyset)\cdots(w_\emptyset - k_\emptyset \pm m_\emptyset - 2(i+j)+2) \cdot (w_\emptyset - k_\emptyset + s_\emptyset - 1 \pm m_\emptyset)\cdots(w_\emptyset - k_\emptyset + s_\emptyset - 2i + 1 \pm m_\emptyset)$, where $i, j \in Z^+$ if $s_\emptyset \geq 1$, $i \in Z^+$, $j = 0$ or 1 if $s_\emptyset = 0$.

Now, let $n_\emptyset = 1$ if $s_\emptyset = 0$, $n_\emptyset = 2$ if $s_\emptyset \geq 1$. Let $1_\emptyset = k_\emptyset - s_\emptyset + 1$, that is, $1_\emptyset = \frac{1}{2}(\dim \underline{g}^\emptyset)$ if $2\emptyset \notin \Delta$, $1_\emptyset = \frac{1}{2}(\dim \underline{g}^\emptyset) + 1$ if $2\emptyset \in \Delta$. Thus $\lambda \in \underline{a}'$ is a root of some $T_{\rho\emptyset - \rho}p^\tau_\delta$, $\delta \in \Gamma_\emptyset$, if and only if
(1) $\langle \lambda + \rho, w_\emptyset \rangle - 1_\emptyset \pm m_\emptyset \in n_\emptyset Z^+$. Now, $(-\lambda - 2\rho) + \rho = -(\lambda + \rho)$, so $-\lambda - 2\rho$ is a root of some $T_{\rho\emptyset - \rho}p^\tau_\delta$ if (2) $-\langle \lambda + \rho, w_\emptyset \rangle - 1_\emptyset \pm m_\emptyset \in n_\emptyset Z^+$. So we get:

Proposition 13.1. There exists γ in Γ such that

(a) $r^\gamma_\tau(\lambda) = 0$, if and only if (1) and (2) occur for some $\emptyset \in \Delta_+^1$.

(b) $p^\tau_\delta(\lambda) = 0$, if and only if (1) occurs for some $\emptyset \in \Delta_+^1$. Now, by Corollary 5.10, we get:

Corollary 13.2.

(a) X^τ_λ is irreducible if and only if, for all $\emptyset \in \Delta_+^1$, (1)' $\langle \lambda + \rho, w_\emptyset \rangle - 1_\emptyset \pm m_\emptyset \notin n_\emptyset Z^+$, and (2)' $-\langle \lambda + \rho, w_\emptyset \rangle - 1_\emptyset \pm m_\emptyset \notin n_\emptyset Z^+$.

(b) X^τ_λ is cyclic if and only if, for all $\emptyset$ in Δ_+^1, (1)' holds.

Remark 13.3. For $\emptyset \in \Delta_+^1$, let $J_\emptyset \subset R$ be the largest connected open interval J containing 0 such that if $t \in J$, $\pm t \pm m_\emptyset - 1_\emptyset \notin n_\emptyset Z^+$. Let CS (= "critical strip") = $\{\lambda \in \underline{a}'$ such that $\mathrm{Re}\langle \lambda + \rho, w_\emptyset \rangle \in J_\emptyset$ for all $\emptyset \in \Delta_+^1\}$. Then if $\lambda \in CS$, X^τ_λ is irreducible. Note that CS = the empty set if and only if there exists $\emptyset$ in Δ_+^1 such that $(\pm m_\emptyset - 1_\emptyset) \in n_\emptyset Z^+$.

Now, recall $R^\gamma_\tau(\lambda) = P^\gamma_\tau(\lambda)P^\gamma_\tau(-\bar{\lambda} - 2\rho)^*$. Thus, if $p^\gamma_\tau(-\lambda - 2\rho) \neq 0$, rank $R^\gamma_\tau(\lambda)$ = rank $P^\gamma_\tau(\lambda)$, and then by Theorem 5.3 and Theorem 5.7 we get $Z^\tau_\lambda = X^{\tau,e}_\lambda$ $(= U \cdot (X^\tau_\lambda)^{\underline{k}^\#})$. Now, by Corollary 13.2(b), $p^\tau_\gamma(\lambda) \neq 0$ if $\mathrm{Re}\langle \lambda + \rho, w_0 \rangle - 1_\emptyset + |m_\emptyset| < 0$ for all $\emptyset \in \Delta_+^1$. Thus, $p^\tau_\gamma(-\lambda - 2\rho) \neq 0$ if $\mathrm{Re}\langle \lambda + \rho, w_\emptyset \rangle > |m_0| - 1_\emptyset$, for all $\emptyset$ in Δ_+^1. So, let $C_0 = \{\lambda \in \underline{a}'$ such that $\mathrm{Re}\langle \lambda = \rho, w_\emptyset \rangle \geq 0$ for all $\emptyset \in \Delta_+^1\}$. C_0 is often called the positive Weyl chamber. Now, if $\lambda \in \underline{a}'$, λ is W_T-conjugate to some element of C_0, and thus by Theorem 4.1, we see that if $\lambda \in \underline{a}'$, $Z^\tau_\lambda = Z^\tau_\nu$ for some ν in C_0. So, we get:

Theorem 13.4.

(a) Suppose that, for all $\emptyset$ in Δ_+^1, $1_\emptyset > |m_\emptyset|$. Then $\{X^{\tau,e}_\lambda$, λ in $C_0\}$ are all irreducible, and these are the only irreducible U-modules Z with $Z^{\underline{k}^\#} \neq 0$.

(b) Suppose there exists $\emptyset$ in Δ_+^1 such that $1_\emptyset \leq |m_\emptyset|$. Then $\{X^{\tau,e}_\lambda$, $\lambda \in C_0$, $\langle \lambda + \rho, w_\emptyset \rangle + 1_\emptyset \pm m_\emptyset \notin n_\emptyset Z^+\}$ are all irreducible, and the only other irreducible U-modules Z such that $Z^{\underline{k}^\#} \neq 0$ are $\{Z^\tau_\lambda$, $\lambda \in C_0$, and such that there exists $\emptyset$ in Δ_+^1, $\langle \lambda + \rho, w_\emptyset \rangle + 1_\emptyset \pm m_\emptyset \in n_\emptyset Z^+\}$, that is, $\{Z^\tau_\lambda$, for λ in the intersection of C_0 with a finite number of hyperplanes in $\underline{a}'\}$.

Remark 13.5. Suppose $m_\emptyset \in \frac{1}{2}n_\emptyset Z$. Then if $\langle \lambda + \rho, w_\emptyset \rangle - 1_\emptyset - |m_\emptyset| \in n_\emptyset Z^+$,

$\langle \lambda + \rho, w_\emptyset \rangle - 1_\emptyset + |m_\emptyset| \in n_\emptyset Z^+$. Thus, if $m_\emptyset \in \frac{1}{2} n_\emptyset Z$, the terms "$\pm m_\emptyset$" in Theorems 13.2 and 13.4 can be replaced by $+|m_\emptyset|$. And, in fact, one can easily see that, if $\underline{g}_R = sl(2,R)$ or $so(n,2)$, and τ is the differential of a character on the maximal compact subgroup of $SL(2,R)$ or $SO(n,2)_c$ (connected component), then $m_\emptyset \in \frac{1}{2}Z = \frac{1}{2}n_\emptyset Z$. If $\underline{g}_R = su(N,1)$, $N \geq 2$, then $m_\emptyset \in Z = \frac{1}{2}n_\emptyset Z$ if τ is the differential of a character on $U(N) \subset SU(N,1)$.

Section 14. Representations of the Group G_0.

Finally we consider an arbitrary connected real semisimple Lie group $G_0 = K_0 A_0 N_0$, with Lie algebra $\underline{g}_R$. Let τ be a character on K_0, and we also write τ for its differential. Recall π_λ^τ is the principal series representation of G_0 on the functions $\{f\colon\ G_0 \to C,\ f(gman) = \tau(m)^{-1} a^{\lambda'} f(g),\ g \text{ in } G_0,\ m \in M_0,\ a \text{ in } A_0,\ n \text{ in } N_0\}$. We are interested in finding the irreducible unitary representations of G_0 with a vector on which K_0 acts according to τ.

Recall $\underline{a}^* = \{\lambda \in \underline{a}'$ such that $(\lambda+\rho) \sim_W -(\overline{\lambda} + \rho)\}$, and that by Theorem 6.1, Z_λ^τ has a U-invariant Hermitian structure if and only if $\lambda \in \underline{a}^*$. Also recall $\underline{a}_+^* = \{\lambda \in \underline{a}^*$ such that Z_λ^τ admits a positive definite U-invariant Hermitian structure$\}$, and that by Theorem 6.4, $\lambda \in \underline{a}_+^*$ if and only if $\lambda \in \underline{a}^*$ and $R_\tau^\gamma(\lambda)$ is positive semi-definite.

Let us now recall Nelson's Theorem (7a.):

Theorem 14.1. Suppose a U-module Z is a Harish-Chandra module, such that there exists u in U, $u = x_1^2 + \ldots + x_{\dim \underline{g}}^2$, where x_i are linearly independent, and u is essentially self-adjoint on Z. Then Z can be completed to a Hilbert space H which is a unitary representation of $G_0{}^s$ (the simply connected covering group of G_0) if and only if Z has a U-invariant positive definite Hermitian structure.

It is easy to see that Z_λ^τ is a Harish-Chandra module. Choosing $u = \omega_{\underline{g}} - 2\omega_{\underline{k}}$ (where ω is the Casimir operator), we have that u is of the form $x_1^2 + \ldots + x_{\dim \underline{g}}^2$, and u acts as a scalar on each $(\tau \otimes Z_\lambda^\tau)_\gamma$, γ in Γ. Thus, u is essentially self-adjoint on Z_λ^τ. Thus Z_λ^τ can be completed to a Hilbert space H_λ^τ on which $G_0{}^s$ is unitary if and only if $\lambda \in \underline{a}_+^*$. Further, by (1a.), since Z_λ^τ is U-irreducible, H_λ^τ is $G_0{}^s$-irreducible. Thus, the center Z^s of G_0^s acts as scalars on H_λ^τ.

But $Z^s \subset K_0^s$, and we know the action of K_0^s on $(Z_\lambda^\tau)^{\underline{k}\#}$ is the character τ^s whose differential is τ. Thus the action of Z^s on Z_λ^τ is τ^s. On the other hand, $G_0 = G_0^s/D$, where $D \subset Z^s$. Also, there exists a character τ_0 on K_0 which has differential τ. Thus we see that τ^s is the identity on D, and so the action of D on H_λ^τ is the identity, and so we have a unitary representation on $G_0^s/D = G_0$.

Now let $\underline{U} = \{\lambda \in \underline{a}'$ such that $\lambda = -\rho + i\nu,\ \nu \in \underline{a}_R'\}$. If $\lambda \in \underline{U}$, we know π_λ^τ is unitary (see section 1), even if it is reducible. Define the complementary series as the set of all irreducible unitary G_0- representations other than π^τ_λ, λ in $\underline{U}$, which have a vector on which K_0 acts according to τ.

Proposition 14.2. If $\lambda \in \underline{a}^* \cap CS$, π^τ_λ is an irreducible unitary representation of G_0. Proof: Note that CS may be the empty set. If not, $-\rho \in CS$, and $X^\tau_{-\rho}$ is irreducible, and thus $= Z^\tau_{-\rho}$, by Corollary 13.3. Thus $R^\gamma_\tau(-\rho)$ is positive and semi-definite (since $Z^\tau_{-\rho}$ can be extended to a unitary G_0-representation $\pi^\tau_{-\rho}$), and in fact is positive definite by the irreducibility of $X^\tau_{-\rho}$.

Now let $\lambda \in \underline{a}^* \in CS$. Then if $0 \leqslant t \leqslant 1$, $-t\rho + (1-t)\lambda \in \underline{a}^* \cap CS$. So, $t \to R^\tau_\gamma\big(-t\rho + (1-t)\lambda\big)$ is a continuous matrix-valued function whose range lies in the non-singular Hermitian matrices $\big(R^\gamma_\tau(\lambda)$ is Hermitian if $\lambda \in \underline{a}^*\big)$. But the positive definite Hermitian matrices form a connected component of the non-singular Hermitian matrices, and thus by continuity, $R^\gamma_\tau(\lambda)$ is positive definite since $R^\gamma_\tau(-\rho)$ is.

Now, X^τ_λ is irreducible, and thus $= Z^\tau_\lambda$, and therefore π^τ_λ is irreducible and unitary.

Q.E.D.

Proposition 14.2 gives a sufficient condition for a principal series representation to be a member of the complementary series. In the rank one case we will exhibit the whole complementary series. Wakimoto (8a.) has found the elements p^τ_γ if $\underline{g}_R = su(1,1)$. If $G_0 = SU(1,1)$, $m = m_\emptyset$ can take on values in $\frac{1}{2}Z$. Further, $\pi^{m\tau_0}_\lambda = \pi^{m'\tau_0}_\lambda$ if $(m-m') \in Z$. Wakimoto proves (a) of the following, and (b) follows from Proposition 14.2.

Theorem 14.3.

(a) Let $m = \frac{1}{2}$. Then the only unitary representations of $G_0 = SU(1,1)$ with a vector on which K_0 acts according to τ are $\{\pi^\tau_{-\rho+i\nu}, \nu \in \underline{a}_R{}', \nu \neq 0\}$, and an irreducible subrepresentation of $\pi^\tau_{-\rho}$. That is, there is no CS, and if $\lambda \in \underline{a}_R{}'$, $\lambda \neq -\rho$, one can find a γ in Γ such that $R^\gamma_\tau(\lambda) = r^\gamma_\tau(\lambda) < 0$.

(b) Let G_0 be any connected group such that $\underline{g}_R = su(1,1)$. Define $m = m_\emptyset$ as before. Then the only irreducible unitary representations of G_0 with a vector on which K_0 acts according to τ are $\pi^{\tau,e}_\lambda$, $\lambda \in \underline{a}^* \cap \overline{CS}$. Here, $\overline{}$ denotes closure. There are $[m+1]^\circ$ additional ones, for $m > 0$.

For the remainder of this section, we assume G_0 is such that $\underline{g}_R = su(N,1)$, $N \geqslant 2$. Recall $w_\emptyset$, $l_\emptyset$, $n_\emptyset$, and $m_\emptyset$ from section 13. Here, $\emptyset$ in $\Delta_+{}^1$ is unique. so, we write w for $w_\emptyset$, m for $m_\emptyset$. We can see that $n_\emptyset = 2$ and $l_\emptyset = N$. We write p^m_γ, X^m_λ, etc., for $p^{m\tau_0}_\gamma$, $X^{m\tau_0}_\lambda$, etc., where, again, τ_0 is the differential of the smallest character on the maximal compact subgroup of $SU(N,1)$. We do not assume that m is an integer. Note, however, that if $G_0 = SU(N,1)$, $m \in Z$, and if $G_0 = Ad\big(SU(N,1)\big)$, $m \in (N+1)Z$. We can assume without loss of generality that $m \geqslant 0$. We will prove:

Theorem 14.4

(a) Suppose $N > m$. Then π_λ^{-m} is unitary and irreducible if and only if $\lambda \in \underline{a}^* \cap CS$. Further, $\pi_{-m\emptyset}^{-m,e}$ $\left(= G_0 \cdot (X_{-m\emptyset}^{-m})^{\underline{k}^\#}\right)$ is irreducible, and these are all irreducible unitary representations of G_0 with a vector on which K_0 acts according to τ_0^{-m}.

(b) Suppose $N \leq m$. CS exists if and only if $m - N \notin 2Z$. Except for $\pi_{-\rho}^{-m}$ (which is irreducible if and only if $m - N \notin 2Z$), π_λ^{-m} is irreducible and unitary if and only if $\lambda \in \underline{a}_R{}'$ or $\lambda \in \underline{U} = \{-\rho + i\nu$ such that $\nu \in \underline{a}_R{}'\}$. There are only $[\frac{1}{2}(m-N)+\frac{3}{2}]°$ additional unitary representations of G_0 with a vector on which K_0 acts according to $\tau_0{}^m$, where [ ]° denotes "greatest integer less than." As U-modules, these are the representations $\{Z_\lambda^m$ such that $\lambda \in \underline{a}_R{}'$, $\langle \lambda + \rho, w\rangle - N + m \in 2Z$, and $0 \leq \langle\lambda + \rho, w\rangle \leq (m - N)\}$. Proof:

(a) In view of Proposition 14.2 and Theorem 13.4(a), we need only show, for all $\lambda \in \underline{a}^* - \overline{CS}$, that there exists γ in Γ such that $r_{-m}^\gamma(\lambda) < 0$, for then we have exhausted all U-modules Z^m with a positive definite U-invariant Hermitian structure.

Now, $\underline{a}^* - \overline{CS} = \{\lambda \in \underline{a}_R{}'$ such that $|\langle\lambda + \rho, w\rangle| > N-m\}$. Recall from section 12 that there exists γ in Γ such that $p_\gamma^m = (w + m)$. So, $r_{-m}^\gamma(\lambda)$ = (up to real positive scalar) $p_\gamma^m(\lambda)\, p_\gamma^m(-\lambda-2\rho) = (\langle\lambda,w\rangle + m)(-\langle\lambda,w\rangle - 2N + m)$. This polynomial has roots $\lambda = -m\emptyset$ and $\lambda = (m-2N)\emptyset$. For $\lambda \in \underline{a}_R{}'$, $r_{-m}^\gamma(\lambda) \to -\infty$ as $\lambda \to \pm\infty$. Thus, for $\lambda \in \underline{a}_R{}'$, $r_{-m}^\gamma(\lambda) < 0$ if $\langle\lambda,w\rangle < (m-2N)$ or $\langle\lambda,w\rangle > -m$, that is, if $\lambda \in \underline{a}_R{}' - \overline{CS} = \underline{a}^* - \overline{CS}$. Finally, $-m\emptyset \in \overline{CS} - CS$.

(b) Since $l_\emptyset = N$, the first statement follows from the definition of CS. By section 1, π_λ^{-m} is unitary if $\lambda \in \underline{U}$, and is irreducible by Corollary 13.2(a), with the exception of $\pi_{-\rho}^{-m}$, which one can see is reducible if and only if CS is empty. Note that if $(m-N) > 0$, but $(m-N) \notin 2Z^+$, the width of CS is between 0 and $2\emptyset$.

Now, $-1 \in W$, so $Z_{\lambda+\rho}^{-m} = Z_{-\lambda+\rho}^{-m}$ for all λ in $\underline{a}'$. So, to prove (b), we must show that, if $\lambda \in \underline{a}_R{}' - CS$, and $\lambda \notin S = \{\lambda | (\langle\lambda + \rho, w\rangle - N + m) \in 2Z$ and $\langle\lambda + \rho, w\rangle \leq (m - N)\}$, then there exists γ in Γ such that $r_{-m}^\gamma(\lambda) < 0$. Further, if $\lambda \in S$, we must show that, for all $\gamma \in \Gamma$, $r_{-m}^\gamma(\lambda) \geq 0$.

By Theorem 12.2, we can index γ in Γ such that $p_{\gamma ij}^m = (w-m)(w-m-2)\cdots(w-m-2i+2)\cdot(w+m)\cdots(w+m-2j+2)$, where i or j > 0. We first prove:

Lemma 14.5. Let $T_j = \{r \in R$ such that $-m+2j < r < -m+2j+2$ or $m-2N-2J-2 < r < m-2N-2j\}$, where $j \in Z^+$, $j \leq \frac{1}{2}(m-N)-1$. Let $\lambda = r\emptyset$, $r \in T_j$. Then $r_{-m}^{\gamma_{0,j+2}}(\lambda) < 0$. Proof: $r_{-m}^{\gamma_{0,k+2}}(\lambda) = \left[(\langle\lambda,w\rangle + m)(\langle-\lambda-2\rho,w\rangle + m)\right]\cdot\left[(\langle\lambda,w\rangle + m-2)(\langle-\lambda-2\rho,w\rangle + m-2)\right]\cdots\left[(\langle\lambda,w\rangle + m-2k-2)\cdot(\langle-\lambda-2\rho,w\rangle + m-2k-2)\right] = \prod_{l=0}^{k+1} a_i(\lambda)$, where $a_i(\lambda) = (\langle\lambda,w\rangle + m-2i)(\langle-\lambda-2\rho,w\rangle + m-2i)$.

But the roots of a_i are $\lambda = (-m+2i)\emptyset$ and $\lambda = (2N+m-2i)\emptyset$, and $a_i(\lambda) \to -\infty$ as $\lambda \to \pm\infty$, λ in $\underline{a}_R{}'$. Thus, if $r \in T_j$, $a_i(r\emptyset) > 0$ for $i = 0,\ldots,j$, while

$a_{j+1}(r\emptyset) < 0$. So, $\prod_{i=0}^{j+1} a_i(r\emptyset) < 0$, and thus $r_{-m}^{\gamma_{0,j+2}}(r\emptyset) < 0$.

Q.E.D.

Calling $T = (\bigcup_{j=0}^{[\frac{m-n}{2}]} T_j) \subset R$, we have shown that if $\lambda \in T\cdot\emptyset - \overline{CS}$, there exists γ in Γ such that $r_{-m}^{\gamma}(\lambda) < 0$. Now, $\underline{a}_R' \cap S \cup T\cdot\emptyset - \overline{CS} \cup CS = \{\lambda$ in $\underline{a}_R'$ such that $|\langle \lambda + \rho, w\rangle| \leq m-N\}$. So, $\underline{a}_R' - (S \cup T\cdot\emptyset - \overline{CS} \cup CS) = \{\lambda$ in $\underline{a}_R'$ such that $\langle \lambda, w\rangle > m-2N$ or $\langle \lambda, w\rangle < -m\}$. But $r_{-m}^{\gamma_{0,1}}$ has roots $\lambda = (m-2N)\emptyset$ and $-m\emptyset$, and so if $\lambda \in \underline{a}_R' - (S \cup T\cdot\emptyset - \overline{CS} \cup CS)$, $r_{-m}^{\gamma_{0,1}}(\lambda) < 0$. Thus if $\lambda \in \underline{a}_R' - CS - S$, there is γ in Γ such that $r_{-m}^{\gamma}(\lambda) < 0$. So, to prove the theorem, we need only show, for all λ in S, for all γ in Γ, $r_{-m}^{\gamma}(\lambda) \geq 0$.

Now, note that if $j,k \geq 1$, $p_{\gamma_{jk}}^m = p_{\gamma_{j0}}^m p_{\gamma_{0k}}^m$, and so $r_{-m}^{\gamma_{jk}} = r_{-m}^{\gamma_{j0}} r_{-m}^{\gamma_{0k}}$. So, to show $r_{-m}^{\gamma_{jk}}(\lambda) \geq 0$ for all λ in S, it suffices to show $r_{-m}^{\gamma_{j0}}(\lambda) \geq 0$ and $r_{-m}^{\gamma_{0k}}(\lambda) \geq 0$, for all λ in S, all j,k in Z^+. Now, $r_{-m}^{\gamma_{j+1,0}}(\lambda) = \big((\langle \lambda, w\rangle - m)(\langle -\lambda-2\rho, w\rangle - m)\big)\big((\langle \lambda, w\rangle - m-2)(\langle -\lambda-2\rho, w\rangle - m-2)\big)\cdots\big((\langle \lambda, w\rangle - m-2j)(\langle -\lambda-2\rho, w\rangle - m-2j)\big) = \prod_{i=0}^{j} b_i(\lambda)$, where $b_i(\lambda) = (\langle \lambda, w\rangle - m-2i)(\langle -\lambda-2\rho, w\rangle - m-2i)$. So, to show $r_{-m}^{\gamma_{j0}}(\lambda) \geq 0$ for all λ in S, it suffices to show $b_i(\lambda) \geq 0$ for all λ in S, i in Z^+. But $b_i(\lambda)$ has roots $\lambda = (m+2i)\emptyset$ and $\lambda = (-m-2N-2i)\emptyset$, and $b_i(\lambda) \to -\infty$ as λ in $\underline{a}_R' \to \pm\infty$. So, $b_i(r\emptyset) \geq 0$ if $-m-2N-2i \leq r \leq m+2i$. But if $r\emptyset \in S$, r lies in this interval.

Now, finally, to show $r_{-m}^{\gamma_{0k}}(\lambda) \geq 0$ for λ in S, suppose $\lambda = (-m+2j)\emptyset$ or $\lambda = (m-2N-2j)\emptyset$. Every λ in S is of this form for some $j \in Z^+$, $0 \leq j \leq [\frac{1}{2}(m+1-N)]^\circ$. Recalling the polynomials $a_i(\lambda)$ from Lemma 14.5, one easily sees $a_i(\lambda) > 0$ for $0 \leq i \leq j-1$, and $a_j(\lambda) = 0$. Thus if $k \geq j+1$, $r_{-m}^{\gamma_{0k}}(\lambda) = \prod_{i=0}^{k-1} a_i(\lambda) = 0$, and if $k \leq j$, $r_{-m}^{0k}(\lambda) > 0$ for λ as specified above. Thus, for each λ in S, and all $j,k \in Z^+$ such that j or $k > 0$, we have shown $r_{-m}^{\gamma_{jk}}(\lambda) \geq 0$.

Q.E.D.

Remark 14.6. Call CI (critical interval) = $CS \cap \underline{a}_R'$. Then we get that, except for a finite number of representations, the complementary series corresponding to τ_0^{-m} consists of π_λ^{-m}, λ in CI. If $N \geq m$, this finite number is 1; if $m \geq N$, the number is $[\frac{1}{2}(m+3-N)]^\circ$.

Section 15. An Application.

Let G = KAN be a real connected semisimple Lie group. We now generalize a method of Wallach (9a., 9b.) to obtain information about the principal series representations π_λ^ξ, for all $\xi \in \hat{M}$, $\lambda \in \underline{a}'$, even if G is not linear.

Let $\xi \in \hat{M}$. We say that ξ is extendible if there exists a finite dimensional irreducible G-module V such that $V_\xi = V^{\overline{N}}$ as an M-module. Wallach proves (Theorem 2.1 in 9a.):

Proposition 15.1. Let G be linear. Then if $\xi \in \hat{M}$, there exists s in W such that $s\cdot\xi$ is extendible.

Recall that $s \cdot \xi$ is defined by choosing a representative k of s. Then $s \cdot \xi(m) = \xi(k^{-1}mk)$ for m in M. Modifying Theorem 4.2 in (9b.) slightly, we get:

Proposition 15.2. If $\xi \in \hat{M}$, and ξ is extendible, then there exists $\lambda \in \underline{a}'$ such that H_λ^ξ (the Hilbert space for π_λ^ξ) has a G-invariant subspace $V_{\xi,\lambda}$ such that $V_{\xi,\lambda}$ is an extension of ξ. In fact, if V is an extension of ξ, $a \in A$, $v \in V^{\overline{N}}$, then $a \cdot v = a^{-\lambda'} v$.

The proof does not require G to be linear. Now, if τ is a character on K, $\xi \in \hat{M}$, define $\tau\xi \in \hat{M}$ by $\tau\xi(m) = \tau(m)\xi(m)$ for m in M. We now prove the following consequence of Proposition 15.1.

Corollary 15.3. Let $\xi \in \hat{M}$. Then there exists a character τ on K, s in W, such that $(\tau)(s \cdot \xi) = s \cdot (\tau\xi)$ is extendible. Proof: Ad (G) is a linear group. Now, $\ker(\mathrm{Ad}) \subset Z_K$, the center of K, and $Z_K \subset M$, since M is the centralizer of A in K. Since ξ is irreducible, and ker Ad $\subset$ center of M, ξ acts as scalars on ker Ad. Thus there exists a character χ on ker Ad such that $\xi|_{\ker \mathrm{Ad}} = \chi$. But since Z_K is abelian, χ can be extended to Z_K (Z_K is a product of lines and tori). And, since K is reductive, $K = K'Z_K$, where K' is the commutator subgroup of K. So, we extend χ to K by $\chi(kz) = \chi(z)$ for k in K', z in Z_K. This is well defined since $K' \cap Z_K = \mathrm{id}$.

Now, let $\tau = \chi^{-1}$. Then $(\tau\xi)|_{\ker \mathrm{Ad}} = \mathrm{id.}$, so $(\tau\xi)$ gives a representation $(\overline{\tau\xi})$ on Ad(M). Thus, there exists s in W such that $s \cdot (\overline{\tau\xi})$ is a representation of Ad(M) extendible to an Ad(G)-module V. But V is a G-module by composing with Ad, and it is clear that $V^{\overline{N}} = V_{s \cdot (\tau\xi)}$ as an M-module. Thus, $s \cdot (\tau\xi)$ is extendible.

Further, pick a representative k of s. Then $s \cdot (\tau\xi)(m) = (\tau\xi)(k^{-1}mk) = \tau(k^{-1}mk)\xi(k^{-1}mk) = \tau(m)\xi(k^{-1}mk) = (\tau)(s \cdot \xi)(m)$, for all m in M.

Q.E.D.

Let $1_\lambda^\tau \in H_\lambda^\tau$, $1_\lambda^\tau(kan) = \tau(k^{-1})a^{\lambda'}$. We prove:

Theorem 15.4. Let $\xi \in \hat{M}$, τ a character on K, $\lambda \in \underline{a}'$, such that $V_{(\tau\xi),\lambda} \subset H_\lambda^{\tau\xi}$ is an extension of $(\tau\xi)$. Let $\nu \in \underline{a}'$. If $1_{\nu-\lambda}^{-\tau}$ is a cyclic vector for $H_{\nu-\lambda}^{-\tau}$, and $0 \neq f \in V_{\xi,\lambda}^{\overline{N}}$, then $1_{\nu-\lambda}^{-\tau} \cdot f$ is a cyclic vector for H_λ^ξ. Proof: We first note that $1_{\nu-\lambda}^{-\tau} f \in H_\nu^\xi$. For, $(1_{\nu-\lambda}^{-\tau} f)(gman) = 1_{\nu-\lambda}^{-\tau}(gman) f(gman) = \left(\tau(m)a^{(\nu-\lambda)'} 1_{\nu-\lambda}^{-\tau}(g)\right)\left(\tau\xi(m^{-1})a^{\lambda'} f(g)\right) = \xi(m^{-1})a^{\nu'}(1_{\nu-\lambda}^{-\tau} f)(g)$.

Now, one knows that $G = \overline{N}AK$. Thus, by our assumption, $G \cdot 1_{\nu-\lambda}^{-\tau} = H_{\nu-\lambda}^{-\tau} = \overline{N}A \cdot 1_{\nu-\lambda}^{-\tau}$ since $K \cdot 1_{\nu-\lambda}^{-\tau} \subset C \cdot 1_{\nu-\lambda}^{-\tau}$. Now, if $f \in V_{\tau\xi,\lambda}^{\overline{N}}$, $\overline{n} \in \overline{N}$, $a \in A$, then $\overline{n}a \cdot f = a^{-\lambda'} \cdot f$; that is, $\overline{N}A \cdot f \subset (C - 0)f$.

So, suppose $H_\nu^{\xi,e}$ is the smallest closed G-invariant subspace of H_ν^ξ containing $1_{\nu-\lambda}^{-\tau} f$. Then, $(G \cdot 1_{\nu-\lambda}^{-\tau})f = (\overline{N}A \cdot 1_{\nu-\lambda}^{-\tau})f = (\overline{N}A \cdot 1_{\nu-\lambda}^{-\tau})(\overline{N}A \cdot f) = \overline{N}A \cdot (1_{\nu-\lambda}^{-\tau} f) \subset G \cdot (1_{\nu-\lambda}^{-\tau} f) \subset H_\nu^{\xi,e}$. Taking derivatives, $\left(U(\underline{g}) \cdot 1_{\nu-\lambda}^{-\tau}\right) f \subset H_\nu^{\xi,e}$, or, taking uniform closures, and restricting functions to K, $C^\infty(K,\tau^{-1}) \cdot f|_K \subset H_\nu^{\xi,e}|_K$, where $C^\infty(K,\tau^{-1}) = \{f\colon K \to C$, such that $f(km) = \tau(m)f(k)\}$.

Thus, $K\cdot C^\infty(K,\tau^{-1})K\cdot f|_K \subset H_\nu^{\xi,e}|_K$, or $C^\infty(K,\tau^{-1})\cdot k\cdot f|_K \subset H_\nu^{\xi,e}|_K$. But $C^\infty(K,\tau^{-1})(K\cdot f)|_K = C^\infty(K;\xi)$ = (by definition) {f smooth: $K \to V_\xi$ such that $f(km) = \xi(m^{-1})f(k)$}. For, $M\cdot f = V_{\tau\xi,\lambda}^{\overline{N}}$, and so $K\cdot f = V_{\tau\xi,\lambda}$. One then uses a partition of unity argument to show that an arbitrary element of $C^\infty(K;\xi)$ is a sum of products of an element in $C^\infty(K,\tau^{-1})$ and an element in $V_{\tau\xi,\lambda}$.

Thus, $C^\infty(K,\xi) \subset H_\nu^{\xi,e}|_K$, and since an element of H_ν^ξ is defined by its values on K, we get $H_\nu^{\xi,e} = H_\nu^\xi$.

Q.E.D.

As a consequence, we obtain:

Theorem 15.5. If $\xi \in \hat{M}$, τ a character on K, λ, $\nu \in \underline{a}'$, are such that $(\tau\xi)$ is extendible, $V_{\tau\xi,\lambda}$ an extension of $(\tau\xi)$, where $V_{\tau\xi,\lambda} \subset H_\lambda^{\tau\xi}$, and the multiplicity of $(\tau\xi)$ in $V_{\tau\xi,\lambda}$ as an M-module is 1, and $1_{\nu-\lambda}^{-\tau}$ and $1_{-2\rho-\overline{\nu}-\lambda}^{-\tau}$ are cyclic vectors for $H_{\nu-\lambda}^{-\tau}$ and $H_{-2\rho-\overline{\nu}-\lambda}^{-\tau}$ respectively, then H_ν^ξ is irreducible.

Proof: Recall (see 9b. or sec. 3) the non-degenerate complex linear pairing between H_ν^ξ and $H_{-2\rho-\overline{\nu}}^\xi$: $(f,g) \to \int_{K/M} \langle f(k),\overline{g(k)}\rangle dk$. Let U be a closed G-invariant subspace of H, $U^\perp$ its annihilator in $H_{-2\rho-\overline{\nu}}^\xi$.

Now, by Frobenius Reciprocity, the multiplicity of $V_{\tau\xi,\lambda}$ in $H_\mu^{\tau\xi} = 1$ for all μ in $\underline{a}'$. Now, if (π,W) is a K-module, let $(\tau \otimes \pi)$ be the K-module with vector space W and K-action $(\tau\otimes\pi)(k)(w) = \tau(k)^{-1}\pi(k)w$. This is the same notation as we used in section 4.

Note that $\tau \otimes H_\mu^\xi = H_\mu^{\tau\xi}$ for all μ in $\underline{a}'$, as a K-module. Thus, the multiplicity of $V_{\tau\xi,\lambda}$ in $\tau \otimes H_\mu^\xi$ as a K-module is 1, especially for $\mu = \nu$ or $-2\rho-\overline{\nu}$. Define T: $V_{\tau\xi,\lambda} \to \tau \otimes H_{\nu-\lambda}^\xi$ by $T(f) = 1_{\nu-\lambda}^{-\tau}f$. Then T is a K-linear injection. Injectivity is obvious, and K-linearity follows from the fact that $1_{\nu-\lambda}^{-\tau}$ is $(\tau \otimes \pi_{\nu-\lambda}^{-\tau})(K)$-fixed. Now we prove:

Lemma 15.6. Either $1_{\nu-\lambda}^{-\tau}\cdot V_{\tau\xi,\lambda}^{\overline{N}} \subset U$ or $1_{-2\rho-\overline{\nu}-\lambda}^{-\tau}V_{\tau\xi,\lambda}^{\overline{N}} \subset U^\perp$. Proof: Let $\tau \otimes H_\nu^\xi = U + U'$ be an orthogonal decomposition of H_ν^ξ into K-modules. Clearly, since $H_\nu^\xi = H_{-2\rho-\overline{\nu}}^\xi$ as a K-module, $U' = U^\perp$ as a K-module.

Now let P and P' be projections from $\tau \otimes H_\nu^\xi$ to U and U' respectively. If $P(1_{\nu-\lambda}^{-\tau}V^{\overline{N}}{}_{\tau\xi,\lambda}) \neq 0 \neq P'(1_{\nu-\lambda}^{-\tau}V^{\overline{N}}{}_{\tau\xi,\lambda})$, then 2 copies of the irreducible M-module $V^{\overline{N}}{}_{\tau\xi,\lambda}$ occur in H_ν^ξ, since P and P' are M-linear. Now, $K\cdot V^{\overline{N}}{}_{\tau\xi,\lambda} = V_{\tau\xi,\lambda}$, so $P(V_{\tau\xi,\lambda})$ and $P'(V_{\tau\xi,\lambda})$ are both K-submodules of $V_{\tau\xi,\lambda}$, and each contains a copy of the uniquely occurring M-module $V^{\overline{N}}{}_{\tau\xi,\lambda}$. Thus, $P(V_{\tau\xi,\lambda}) \oplus P'(V_{\tau\xi,\lambda})$ is not an abstract K-submodule of $V_{\tau\xi,\lambda}$, so some irreducible K-component of $V_{\tau\xi,\lambda}$ occurs twice in H_ν^ξ, contradicting the fact that the multiplicity of $V_{\tau\xi,\lambda}$ in H_ν^ξ as a K-module was 1. So, either $P(1_{\nu-\lambda}^{-\tau}V^{\overline{N}}{}_{\tau\xi,\lambda}) = 0$ or $P'(1_{\nu-\lambda}^{-\tau}V^{\overline{N}}{}_{\tau\xi,\lambda}) = 0$. If the latter, $1_{\nu-\lambda}^{-\tau}V^{\overline{N}}{}_{\tau\xi,\lambda} \subset \tau \otimes U$, and we have proved the the claim. If the former (recalling $U' = U^\perp$ as a K-module), we have a copy of $V^{\overline{N}}{}_{\tau\xi,\lambda}$ as an M-module in $\tau \otimes U^\perp$. But $1_{-2\rho-\overline{\nu}-\lambda}^{-\tau}V^{\overline{N}}{}_{\tau\xi,\lambda}$ is a copy of $V^{\overline{N}}{}_{\tau\xi,\lambda}$

in $\tau \otimes H^{\xi}_{-2\rho-\overline{\nu}}$. We use the same argument to find a contradiction, unless $1^{-\tau}_{-2\rho-\overline{\nu}}V^{\overline{N}}{}_{\tau\xi,\lambda} \subset \tau \otimes U^{\perp}$ as an M-module. Thus we have the lemma.

Now, recall Theorem 15.4. Either $G\cdot(1^{-\tau}_{\nu-\lambda}V^{\overline{N}}{}_{\tau\xi,\lambda}) = H^{\xi}_{\nu}$ or $G\cdot(1^{-\tau}_{-2\rho-\overline{\nu}-\lambda}V^{\overline{N}}{}_{\tau\xi,\lambda}) = H^{\xi}_{-2\rho-\overline{\nu}}$, that is, either $U = H$ or $U^{\perp} = H^{\xi}_{-2\rho-\overline{\nu}}$, and thus $U = 0$. So, H^{ξ}_{ν} is irreducible. This proves Theorem 15.5.

Q.E.D.

Bibliography

1a. Harish-Chandra. "Representations of a semisimple Lie group on a Banach space I", Trans. Amer. Math. Soc., vol. 75 (1953), pp. 185-243.

1b. Harish-Chandra. "Spherical functions on a semisimple Lie group I", Amer. Jour. Math., vol. 80 (1958), pp. 241-310.

2a. S. Helgason. Differential Geometry and Symmetric Spaces, Academic Press, New York, 1962.

2b. S. Helgason. "A duality for symmetric spaces with applications to group representationa", Advances in Math., vol. 5 (1970), pp. 1-154.

3a. J. Humphreys. Introduction to Lie Algebras and Representation Theory, Springer, New York and Berlin, 1972, Graduate Texts in Mathematics, vol. 9.

4a. B. Kostant. "Irreducibility of certain series of representations", Publication of 1971 Summer School in Math., I. M. Gel'fand, editor, Bolyai-Janos Math. Soc., Budapest.

5a. B. Kostant and S. Rallis. "Orbits and representations associated with symmetric spaces", Amer. Jour. Math., vol. 93 (1971), pp. 753-809.

6a. A. Magnus. "Principal series of a semisimple Lie group", Master's Thesis, Mass. Inst. of Tech., May, 1976.

7a. E. Nelson. "Analytic Vectors", Annals of Math., (2), vol. 70 (1969), pp. 562-615.

8a. M. Wakimoto. "On the irreducibility of some series of representations", Hiroshima Math. Jour., vol. 2 (1972), pp. 71-98.

8b. M. Wakimoto. "On the irreducibility of induced representations of SU(2,1)", Hiroshima Math. Jour., vol. 3 (1973), pp. 391-406.

9a. N. Wallach. "Cyclic vectors and irreducibility for principal series representations", *Trans. Amer. Math. Soc.*, vol. 158 (1971), pp. 107-113.

9b. N. Wallach. "Cyclic vectors and irreducibility for principal series representations II", *Trans. Amer. Math. Soc.*, vol. 164 (1972), pp. 389-396.

9c. N. Wallach. *Harmonic Analysis and Homogeneous Spaces*, Dekker, N.Y., 1974.

The author holds the position of Senior Engineer in the Aerodynamics Research unit of the Boeing Company.